PENGUIN BOOKS

General Relativity

Leonard Susskind has been the Felix Bloch Professor in Theoretical
Physics at Stanford University since 1978, and his online lectures are
viewed all around the world. One of the fathers of string theory, he is
the author of *The Black Hole War* and *The Cosmic Landscape*.

André Cabannes taught mathematics at the Massachusetts Institute
of Technology and translated the Theoretical Minimum series into
French. He lives in France.

T0200258

This book is the fourth volume of The Theoretical Minimum series. The first volume, *The Theoretical Minimum: What You Need to Know to Start Doing Physics*, covered classical mechanics, which is the core of any physics education. We will refer to it from time to time simply as volume 1. The second book, volume 2, explains quantum mechanics and its relationship to classical mechanics. Volume 3 covers special relativity and classical field theory. This fourth volume expands on that to explore general relativity.

GENERAL RELATIVITY

THE THEORETICAL MINIMUM

LEONARD SUSSKIND
& ANDRÉ CABANNES

PENGUIN BOOKS

PENGUIN BOOKS

UK | USA | Canada | Ireland | Australia
India | New Zealand | South Africa

Penguin Books is part of the Penguin Random House group of companies
whose addresses can be found at global.penguinrandomhouse.com.

First published in the United States of America by Basic Books,
Hachette Book Group 2023
First published in Great Britain by Allen Lane 2023
Published in Penguin Books 2024

005

Printed and bound in Great Britain by Clays Ltd, Elcograf S.p.A.

The authorized representative in the EEA is Penguin Random House Ireland,
Morrison Chambers, 32 Nassau Street, Dublin D02 YH68

A CIP catalogue record for this book is available from the British Library

ISBN: 978–0–141–99986–9

www.greenpenguin.co.uk

Penguin Random House is committed to a
sustainable future for our business, our readers
and our planet. This book is made from Forest
Stewardship Council® certified paper.

To my family
—LS

To my parents,
who taught me work and tenacity
—AC

Contents

Preface

This fourth volume in The Theoretical Minimum (TTM) series on general relativity is the natural continuation of the third volume on special relativity.

In special relativity, Einstein, starting from a very simple principle – the laws of physics should be the same in indistinguishable Galilean referentials – deeply clarified in a couple of papers published in 1905 the various disturbing observations physicists had made and the equations they had written in the last years of the nineteenth and the first years of the twentieth century concerning light and other phenomena.

Special relativity led to a strange description of space-time where time and space were inextricably mingled. For instance, it explained how particles whose lifetime is measured in fractions of a second can have, in our referential, a travel time from the Sun to Earth of more than eight minutes.

Then, from 1907 until 1915, essentially alone, Einstein reproduced his feat starting now from another very simple principle – acceleration and uniform gravity are equivalent. He generalized special relativity to a space-time containing massive bodies. The theory is called *general relativity* (GR). It led to an even stranger description of space-time where masses bend light and more generally warp space and time.

In lecture 1, we prepare the groundwork. We show how the equivalence principle inescapably leads to the bending of light rays by massive bodies.

Lecture 2 is devoted to tensor mathematics because in GR we must frequently change referentials and the equations relating coordinates in one referential to coordinates in another are tensor equations.

Then a large part of the theory is expressed using tensor equations because they have the great quality that if they hold in one referential, they hold in all of them.

Lectures 3, 4, and 5 are devoted to the geometry of Riemannian space and Minkowskian space-time because it can be said, very summarily, that gravity is geometry in a Minkowskian space-time.

In lectures 6, 7, and 8, we explore black holes, not so much because they are interesting astronomical phenomena per se, than because they are the equivalent in Minkowskian space-time of point masses in Newtonian mechanics. Space-time however presents a stranger behavior in the vicinity of a black hole than Newtonian space in the vicinity of a point mass. Understanding well black holes, the *metric* they create, their horizon, time and gravity in the vicinity of their horizon, the way people in and out of a black hole can communicate, etc. is a prerequisite to understanding GR.

In lecture 9 we sketch the derivation of Einstein field equations. And in lecture 10 we present a simple application predicting gravity waves.

This book, as the preceding ones in the series, is adapted from a course I gave for several years, with much pleasure, at Stanford in the Continuing Studies program to an audience of adults.

My coauthor this time is André Cabannes. Even though he is not a professional scientist, his scientific training, including a Stanford doctorate and a couple of years of teaching applied mathematics at the Massachusetts Institute of Technology (MIT), helped him assist me.

May Einstein's way of doing physics – starting from the simplest principles and pursuing dauntlessly the mathematics and the physics to their ultimate consequences, however unsettling they may be – as I have strived to show in this book, be a source of inspiration to young and future physicists.

Leonard Susskind
Palo Alto, California
Fall 2022

Ten years ago, when two of my children, then in their late teens, were studying sciences to enter the French system of grandes écoles, I decided to brush up what I had learned in the seventies in order to accompany them in their studies. I discovered that the Internet had profoundly changed the learning landscape. Beside reading books, one could now also take excellent free courses on the Net. I leisurely attended courses in mathematics, physics, computer science, etc. from MIT, Stanford, and other places. The subject matters often were better explained, the courses more lively and easier to understand, than what I had experienced in the past. One could choose courses by the world's best teachers.

Among these courses was The Theoretical Minimum series by Leonard Susskind, famous among other reasons for his pioneering work on string theory. I liked them so much that when I discovered that two of his filmed physics courses had already been transformed into books, I decided to translate them in French. Later I also translated the third book. Then, since the next volume didn't exist in English yet, I took up writing the English notes as well, having in mind that this work might turn out to be useful. After a lot more work with Professor Susskind and Basic Books team, volume 4 in The Theoretical Minimum series, on general relativity, that you hold in your hands is the result.

I belong to the group of people to whom these so-called Continuing Studies courses were intended: individuals who studied physics at the undergraduate and sometimes graduate level when they were students, then did other things in life, but kept an interest in sciences and would like to have some exposure to where physics stands today at a level above plain vulgarization. Indeed, personally, I have always found vulgarization more confusing and harder to understand than real explanations with some equations.

Leonard's courses gave me access to Lagrangian classical mechanics, quantum mechanics, and classical field theory with a clarity that I had never known before. With his pedagogy and presentation it becomes a pleasure to learn. Of course, it is all the more true when there is no examination of any sort at the end. But the courses and books turned out to be useful for students as well, to prepare for more advanced and academic studies.

So whether you are someone who only wants to have some real understanding of what general relativity is about – the stuff on gravitation that is geometry, masses that bend space, light, and time, black holes out there that you should avoid falling into, gravity waves that we begin to detect, etc. – or you are a student in physics who wants to have a first presentation of general relativity, this book is for you.

André Cabannes
Saint-Cyr-sur-mer,
French Riviera
Fall 2022

Lecture 1: Equivalence Principle
and Tensor Analysis

Andy: *So if I am in an elevator and I feel really heavy, I can't know whether the elevator is accelerating or you mischievously put me on Jupiter?*

Lenny: *That's right, you can't.*

Andy: *But, at least on Jupiter, if I keep still, light rays won't bend.*

Lenny: *Oh yes they will.*

Andy: *Hmm, I see.*

Lenny: *And if you are falling into a black hole, beware, things will get really strange. But, don't worry, I'll shed some light on this.*

Andy: *Er, bent or straight?*

Introduction
Equivalence principle
Accelerated reference frames
Curvilinear coordinate transformations
Effect of gravity on light
Tidal forces
Non-Euclidean geometry
Riemannian geometry
Metric tensor
Mathematical interlude: Dummy variables
Mathematical interlude: Einstein summation convention
First tensor rule: Contravariant components of vectors
Mathematical interlude: Vectors and tensors
Second tensor rule: Covariant components of vectors
Covariant and contravariant components of vectors and tensors

Introduction

General Relativity is the fourth volume in The Theoretical Minimum (TTM) series. The first three were devoted respectively to classical mechanics, quantum mechanics, and special relativity and classical field theory. The first volume laid out the Lagrangian and Hamiltonian description of physical phenomena and the principle of least action, which is one of the fundamental principles underlying all of physics (see volume 3, lecture 7 on fundamental principles and gauge invariance). They were used in the first three volumes and will continue in this and subsequent ones.

Physics extensively uses mathematics as its toolbox to construct formal, quantifiable, workable theories of natural phenomena. The main tools we used so far are trigonometry, vector spaces, and calculus, that is, differentiation and integration. They have been explained in volume 1 as well as in brief refresher sections in the other volumes. We assume that the reader is familiar with these mathematical tools and with the physical ideas presented in volumes 1 and 3. The present volume 4, like volumes 1 and 3 (but unlike volume 2), deals with classical physics in the sense that no quantum uncertainty is involved.

We also began to make light use of tensors in volume 3 on special relativity and classical field theory. Now with general relativity we are going to use them extensively. We shall study them in detail. As the reader remembers, tensors generalize vectors. Just as vectors have different representations, with different sets of numbers (components of the vector) depending on the basis used to chart the vector space they form, this is true of tensors as well. The same tensor will have different components in different coordinate systems. The rules to go from one set of components to another will play a fundamental role. Moreover, we will work mostly with *tensor fields*, which are sets of tensors, a different tensor attached to each point of a space. Tensors were invented by Ricci-Curbastro and Levi-Civita[1] to develop work of Gauss[2]

---

[1] Gregorio Ricci-Curbastro (1853–1925) and his student Tullio Levi-Civita (1873–1941) were Italian mathematicians. Their most important joint paper is "Méthodes de calcul différentiel absolu et leurs applications," in *Mathematische Annalen* 54 (1900), pp. 125–201. They did not use the word *tensor*, which was introduced later by other people.

[2] Carl Friedrich Gauss (1777–1855), German mathematician.

on curvature of surfaces and Riemann[3] on non-Euclidean geometry. Einstein[4] made extensive use of tensors to build his theory of general relativity. He also made important contributions to their usage: the standard notation for indices and the Einstein summation convention.

In *Savants et écrivains* (1910), Poincaré[5] writes that "in mathematical sciences, a good notation has the same philosophical importance as a good classification in natural sciences." In this book we will take care to always use the clearest and lightest notation possible.

Equivalence Principle

Einstein's revolutionary papers of 1905 on special relativity deeply clarified and extended ideas that several other physicists and mathematicians – Lorentz,[6] Poincaré, and others – had been working on for a few years. Einstein investigated the consequences of the fact that the laws of physics, in particular the behavior of light, are the same in different inertial reference frames. He deduced from that a new explanation of the Lorentz transformations, of the relativity of time, of the equivalence of mass and energy, etc.

After 1905, Einstein began to think about extending the principle of relativity to any kind of reference frames, frames that may be accelerating with respect to one another, not just inertial frames. An *inertial frame* is one where Newton's laws, relating forces and motions, have simple expressions. Or, if you prefer a more vivid image, and you know how to juggle, it is a frame of reference in which you can juggle with no problem – for instance in a railway car moving uniformly, without jerks or accelerations of any sort. After ten years of efforts to build a theory extending the principle of relativity to frames with acceleration and taking into account gravitation in a novel way, Einstein published his work in November 1915. Unlike special relativity, which topped off the work of many, general relativity is essentially the work of one man.

---

[3]Bernhard Riemann (1826–1866), German mathematician.
[4]Albert Einstein (1879–1955), German, Swiss, German again, and finally American physicist.
[5]Henri Poincaré (1854–1912), French mathematician.
[6]Hendrik Antoon Lorentz (1853–1928), Dutch physicist.

We shall start our study of general relativity pretty much where Einstein started. It was a pattern in Einstein's thinking to start with a really simple elementary fact, which almost a child could understand, and deduce these incredibly far-reaching consequences. We think that it is also the best way to teach it, to start with the simplest things and deduce the consequences.

So we shall begin with the *equivalence principle*. What is the equivalence principle? It is the principle that says that *gravity is in some sense the same thing as acceleration*. We shall explain precisely what is meant by that, and give examples of how Einstein used it. From there, we shall ask ourselves: what kind of mathematical structure must a theory have for the equivalence principle to be true? What kind of mathematics must we use to describe it?

Most readers have probably heard that general relativity is a theory not only about gravity but also about geometry. So it is interesting to start at the beginning and ask what is it that led Einstein to say that gravity has something to do with geometry. What does it mean to say that "gravity equals acceleration"? You all know that if you are in an accelerated frame of reference, say, an elevator accelerating upward or downward, you feel an effective gravitational field. Children know this because they feel it.

What follows may be overkill, but making some mathematics out of the motion of an elevator is useful to see in a very simple example how physicists transform a natural phenomenon into mathematics, and then to see how the mathematics is used to make predictions about the phenomenon.

Before proceeding, let's stress that the following study on an elevator, and the laws of physics as perceived inside it, is simple. Yet it is a first presentation of very important concepts. It is fundamental to understand it very well. Indeed, we will often refer to it. In lectures 4 to 9, it will strongly help us understand acceleration, gravitation, and how gravitation "warps" space-time.

So let's imagine the Einstein thought experiment where somebody is in an elevator; see figure 1. In later textbooks, it got promoted to a rocket ship. But I have never been in a rocket ship, whereas

I have been in an elevator. So I know what it feels like when it accelerates or decelerates. Let's say that the elevator is moving upward with a velocity v.

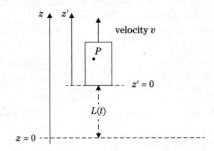

Figure 1: Elevator and two reference frames.

So far the problem is one-dimensional. We are only interested in the vertical direction. There are two reference frames: one is fixed with respect to Earth. It uses the coordinate z. The other is fixed with respect to the elevator. It uses the coordinate z'. A point P anywhere along the vertical axis has two coordinates: coordinate z in the stationary frame, and coordinate z' in the elevator frame. For instance, the floor of the elevator has coordinate $z' = 0$. Its z-coordinate is the distance L, which is obviously a function of time. So we can write for any point P

$$z' = z - L(t) \qquad (1)$$

We are going to be interested in the following question: if we know the laws of physics in the frame z, what are they in the frame z'?

One warning about this lecture: at least at the start, we are going to ignore special relativity. This is tantamount to saying that we are pretending that the speed of light is infinite, or that we are talking about motions so slow that the speed of light can be regarded as infinitely fast. You might wonder: if general relativity is the generalization of special relativity, how did Einstein manage to start thinking about general relativity without including special relativity?

The answer is that special relativity has to do with very high velocities, while gravity has to do with heavy masses. There is a range of situations where gravity is important but high velocities are not. So Einstein started out thinking about gravity for slow velocities, and only later combined it with special relativity to think about the combination of fast velocities and gravity. And that became the general theory.

Let's see what we know for slow velocities. Suppose that z' and z are both inertial reference frames. That means, among other things, that they are related by uniform velocity:

$$L(t) = vt \tag{2}$$

We have chosen the coordinates such that when $t = 0$, they line up. At $t = 0$, for any point, z and z' are equal. For instance, at $t = 0$ the elevator's floor has coordinate 0 in both frames. Then the floor starts rising, its height z equaling vt. So for any point we can write equation (1). In view of equation (2), it becomes

$$z' = z - vt \tag{3}$$

Notice that this is a *coordinate transformation* involving space and time. For readers who are familiar with volume 3 of TTM on special relativity, this naturally raises the question: what about time in the reference frame of the elevator? If we are going to forget special relativity, then we can just say that t' and t are the same thing. We don't have to think about Lorentz transformations and their consequences. So the other half of the coordinate transformation would be $t' = t$.

We could also add to the stationary frame a coordinate x going horizontally and a coordinate y jutting out of the page. Correspondingly, coordinates x' and y' could be attached to the elevator; see figure 2. The x-coordinate will play a role in a moment with a light beam. As long as the elevator is not sliding horizontally, x' and x can be taken to be equal. Same for y' and y.

For the sake of clarity of the drawing in figure 2, we offset a bit the elevator to the right of the z-axis. But think of the two vertical axes as actually sliding on each other, and at $t = 0$ the two origins O and O' coincide. Once again, the elevator moves only vertically.

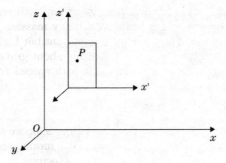

Figure 2: Elevator and two reference frames, three axes in each case.

Finally our complete coordinate transformation is

$$z' = z - vt$$
$$t' = t$$
$$x' = x \tag{4}$$
$$y' = y$$

It is a coordinate transformation of space-time coordinates. For any point P in space-time, it expresses its coordinates in the moving reference frame of the elevator as functions of its coordinates in the stationary frame. It is rather trivial. Only one coordinate, namely z, is involved in an interesting way.

Let us look at a law of physics expressed in the stationary frame. Take Newton's law of motion $F = ma$ applied to an object or a particle. The acceleration a is $\ddot{z}$, where z is the vertical coordinate of the particle. So we can write

$$F = m\ddot{z} \tag{5}$$

As we know, $\ddot{z}$ is the second time derivative of z with respect to time – it is called the vertical acceleration – and F of course is the vertical component of force. The other components we will take to be zero. Whatever force is exerted, it is exerted vertically. What could this force be due to? It could be related to the elevator or not. There could be some charge in the elevator pushing on the particle. Or it could just be a force due to a rope

attached to the ceiling and to the particle that pulls on it. There
could be a field force along the vertical axis. Any kind of force
could be acting on the particle. Whatever the causes, we know
from Newton's law that the equation of motion of the particle, expressed in the original frame of reference, is given by equation (5).

What is the equation of motion expressed in the primed frame?
This is very easy. All we have to do is figure out what the original
acceleration is in terms of the primed acceleration. What is the
primed acceleration? It is the second derivative with respect to
time of z'. Using the first equation in equations (4)

$$z' = z - vt$$

one differentiation gives

$$\dot{z}' = \dot{z} - v$$

and a second one gives

$$\ddot{z}' = \ddot{z}$$

The accelerations in the two frames of reference are the same.

All this should be familiar. But I want to formalize it to bring out
some points. In particular, I want to stress that *we are doing a
coordinate transformation*. We are asking how the laws of physics
change in going from one frame to another. What can we now
say about Newton's law in the primed frame of reference? We
substitute $\ddot{z}'$ for $\ddot{z}$ in equation (5). As they are equal, we get

$$F = m\ddot{z}' \qquad\qquad (6)$$

We found that Newton's law in the primed frame is exactly the
same as Newton's law in the unprimed frame. That is not surprising. The two frames of reference are moving with uniform
velocity relative to each other. If one of them is an inertial frame,
the other is an inertial frame. Newton taught us that the laws of
physics are the same in all inertial frames. It is sometimes called
the *Galilean principle of relativity*. We just formalized it.

Let's turn to an accelerated reference frame.

Accelerated Reference Frames

Suppose that $L(t)$ from figure 1 is increasing in an accelerated way. The height of the elevator's floor is now given by

$$L(t) = \frac{1}{2}gt^2 \tag{7}$$

We use the letter g for the acceleration because we will discover that the acceleration mimics a gravitational field – as we feel when we take an elevator and it accelerates. We know from volume 1 of TTM on classical mechanics or from high school, that this is a uniform acceleration. Indeed, if we differentiate $L(t)$ with respect to time, after one differentiation we get

$$\dot{L} = gt$$

which means that the velocity of the elevator increases linearly with time. After a second differentiation with respect to time, we get

$$\ddot{L} = g$$

This means that the acceleration of the elevator is constant. The elevator is uniformly accelerated upward. The equations connecting the primed and unprimed coordinates are different from equations (4). The transformation for the vertical coordinates is now

$$z' = z - \frac{1}{2}gt^2 \tag{8}$$

The other equations in equations (4) don't change:

$$t' = t$$
$$x' = x$$
$$y' = y$$

These four equations are our new coordinate transformation to represent the relationship between coordinates that are accelerated relative to each other.

We will continue to assume that in the z, or unprimed, coordinate system, the laws of physics are exactly what Newton taught us. In other words, the stationary reference frame is inertial, and we have $F = m\ddot{z}$. But the primed frame is no longer inertial. It is in uniform acceleration relative to the unprimed frame. Let's ask what the laws of physics are now in the primed frame of reference. We have to do the operation of differentiating twice over again on equation (8). We know the answer:

$$\ddot{z}' = \ddot{z} - g \tag{9}$$

Ah ha! Now the primed acceleration and the unprimed acceleration differ by an amount g. To write Newton's equations in the primed frame of reference, we multiply both sides of equation (9) by m, the particle mass, and we replace $m\ddot{z}$ by F. We get

$$m\ddot{z}' = F - mg \tag{10}$$

We have arrived at what we wanted. Equation (10) looks like a Newton equation, that is, mass times acceleration is equal to some term. That term, $F - mg$, we call the force in the primed frame of reference. You notice, as expected, that the force in the primed frame of reference has an extra term: the mass of the particle times the acceleration of the elevator, with a minus sign.

What is interesting about the "fictitious force" $-mg$, in equation (10), is that it looks exactly like the force exerted on the particle by gravity on the surface of the Earth or the surface of any kind of large massive body. That is why we called the acceleration g. The letter g stood for gravity. It looks like a uniform gravitational field. Let me spell out in what sense it looks like gravity. The special feature of gravity is that gravitational forces are proportional to mass – the same mass that appears in Newton's equation of motion. We sometimes say that *the gravitational mass is the same as the inertial mass*. That has deep implications. If the equation of motion is

$$F = ma \tag{11}$$

and the force itself is proportional to mass, then the mass cancels in equation (11). That is a characteristic of gravitational forces: for a small object moving in a gravitational force field, its motion doesn't depend on its mass. An example is the motion of the Earth about the Sun. It is independent of the mass of the Earth. If you know where the Earth is at time t, and you know its velocity at that time, then you can predict its trajectory. You don't need to know what the Earth's mass is.

Equation (10) is an example of *fictitious force* – if you want to call it that – mimicking the effect of gravity. Most people before Einstein considered this largely an accident. They certainly knew that the effect of acceleration mimics the effect of gravity, but they didn't pay much attention to it. It was Einstein who said: look, this is a deep principle of nature that gravitational forces cannot be distinguished from the effect of an accelerated reference frame.

If you are in an elevator without windows and you feel that your body has some weight, you cannot say whether the elevator, with you inside, is resting on the surface of a planet or, far away from any massive body in the universe, some impish devil is accelerating your elevator. That is the *equivalence principle*. It extends the relativity principle, which said you can juggle in the same way at rest or in a railway car in uniform motion. With a simple example, we have equated accelerated motion and gravity. We have begun to explain what is meant by the sentence: "gravity is in some sense the same thing as acceleration."

We have to discuss this result a bit, though. Do we really believe it totally or does it have to be qualified? Before we do that, let's draw some pictures of what these various coordinate transformations look like.

Curvilinear Coordinate Transformations

Let's first consider the case where $L(t)$ is proportional to t. That is when we have

$$z' = z - vt$$

In figure 3, every point – also called *event* – in space-time has a pair of coordinates z and t in the stationary frame and also a pair of coordinates z' and t' in the elevator frame. Of course, $t' = t$ and we left out the two other spatial coordinates x and y, which don't change between the stationary frame and the elevator. We represented the time trajectories of fixed z with dotted lines and of fixed z' with solid lines.

A fundamental idea to grasp is that events in space-time exist irrespective of their coordinates, just as points in space don't depend
on the map we use. Coordinates are just some sort of convenient
tags. We can use whichever we like. We'll stress it again after we
have looked at figures 3 and 4.

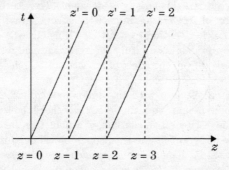

Figure 3: Linear coordinate transformation. The coordinates (z', t')
are represented in the basic coordinates (z, t). An event is a point on
the page. It has one set of coordinates in the (z, t) frame and another
set in the (z', t') frame. Here the transformation is simple and linear.

That is called a *linear coordinate transformation* between the two
frames of reference. Straight lines go to straight lines, not surprisingly since Newton tells us that free particles move in straight
lines in an inertial frame of reference. What is a straight line
in one frame had therefore better be a straight line in the other
frame. Not only do free particles move in straight lines in space,
when we add x and y, but their trajectories are straight lines in
space-time – straight in space and with uniform velocity.

Let's do the same thing for the accelerated coordinate system. The
transformation equation is now equation (8) linking z' and z. The
other coordinates don't change. Again, in figure 4, every point in
space-time has two pairs of coordinates (z, t) and (z', t'). The
time trajectories of fixed z, represented with dotted lines, don't
change. But now the time trajectories of fixed z' are parabolas
lying on their side. We can even represent negative times in the
past. Think of the elevator that was initially moving downward
with a negative velocity but a positive acceleration g (in other
words, slowing down). Then the elevator bounces back upward

with the same acceleration g. Each parabola is just shifted relative to the previous one by one unit to the right.

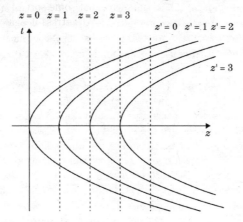

Figure 4: Curvilinear coordinate transformation.

What figure 4 illustrates is, not surprisingly, that straight lines in one frame are not straight lines in the other frame. They become curved lines. As regards the lines of fixed t or fixed t', they are of course the same horizontal straight lines in both frames. We haven't represented them.

We should view figure 4 as just two sets of coordinates to locate each point in space-time. One set of coordinates has straight axes, while the second – represented in the first frame – is curvilinear. Its lines $z' = $ constant are actually curves, while its lines $t' = $ constant are horizontal straight lines. So it is a *curvilinear coordinate transformation*.

Let's insist on the way to interpret and use figure 4 because it is fundamental to understand it very well if we want to understand the theory of relativity – special relativity and even more importantly general relativity. The page represents space-time – here, one spatial dimension and one temporal dimension.

Points (= events) in space-time are points on the page. An event *does not have two positions* on the page, i.e., in space-time. *It has only one position* on the page. But this position can be located, mapped,"charted"one also says, using several differentsystemsof reference. A system of reference, also called a frame of reference, is nothing more than a complete set of "labels," if you will, attaching one label (consisting of two numbers, because our space-time here is two-dimensional) to each point, i.e., to each event.

In a two-dimensional space, the system of reference can be geometrically simple, like orthogonal Cartesian axes in the plane. However this is not a necessity. For one thing, on Earth, which is not a plane, the axes are not straight lines. The usual axes used by cartographers and mariners are meridians and parallels. But on a 2D surface, be it a plane or not, we can imagine quite fancy or intricate curvilinear lines to serve as a frame of reference – so long as it attaches unequivocally two numbers to each (by definition, fixed) point. This is what figure 4 does in the space-time made of one temporal and one spatial dimension represented on the page. We will see many more in lecture 2.

Something Einstein understood very early is this:

There is a connection between gravity and curvilinear coordinate transformations of space-time.

Special relativity was only about linear transformations – transformations that take uniform velocity to uniform velocity. Lorentz transformations are of that nature. They take straight lines in space-time to straight lines in space-time. However, if we want to mock up gravitational fields with the effect of acceleration, we are really talking about transformations of coordinates of space-time that are curvilinear. That sounds extremely trivial. When Einstein said it, probably every physicist knew it and thought: "Oh yeah, no big deal." But Einstein was very clever and very persistent. He realized that if he followed very far the consequences of this, he could then answer questions that nobody knew how to answer.

Let's look at a simple example of a question that Einstein answered using the curved coordinates of space-time representing acceleration, and consequently, if the two are the same, gravity. The question is: what is the influence of gravity on light?

Effect of Gravity on Light

When Einstein first asked himself the question "what is the influence of gravity on light"? around 1907, most physicists would have answered: "There is no effect of gravity on light. Light is light. Gravity is gravity. A light wave moving near a massive object moves in a straight line. It is a law of light that it moves in straight lines. And there is no reason to think that gravity has any effect on it."

But Einstein said: "No, if this equivalence principle between acceleration and gravity is true, then gravity must affect light. Why? Because acceleration affects light." It was again one of these arguments that you could explain to a clever child.

Let's imagine that, at $t = 0$, a flashlight (today we might use a laser pointer) emits a pulse of light in a horizontal direction from the left side of the elevator; see figure 5. The light then travels across to the right side with the usual speed of light c. Since the stationary frame is assumed to be an inertial frame, the light moves in a straight line in the stationary frame.

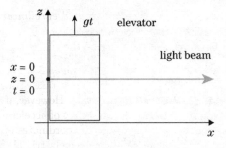

Figure 5: Trajectory of a light beam in the *stationary* reference frame.

The equations for the light ray are

$$x = ct$$
$$z = 0 \qquad (12)$$

The first of these equations just says that the light moves across the elevator with the speed of light – no surprise here.

The second says that in the stationary frame the trajectory of the light beam is horizontal.

Let's express the same equations in terms of the primed coordinates. The first equation becomes

$$x' = ct$$

And the second takes the more interesting form

$$z' = -\frac{g}{2}t^2$$

It says that as the light ray moves across the elevator, at the same time the light ray accelerates downward – toward the floor – just as if gravity were pulling it.

We can even eliminate t from the two equations and get an equation for the curved trajectory of the light ray:

$$z' = -\frac{g}{2c^2}x'^2 \tag{13}$$

Thus, the trajectory, in the primed frame of reference, is a parabola, not a straight line.

But, said Einstein, if the effect of acceleration is to bend the trajectory of a light ray, then so must be the effect of gravity.

Andy: *Gee Lenny, that's really simple. Is that all there is to it?*

Lenny: *Yup Andy, that's all there is to it. And you can bet that a lot of physicists were kicking themselves for not thinking of it.*

To summarize, in the stationary frame, the photon trajectory (figure 5) is a straight line, while in the elevator reference frame, it is a parabola (figure 6).

Let's imagine three people arguing. I am in the elevator, and I say: "Gravity is pulling the light beam down." You are in the stationary frame, and you say: "No, it's just that the elevator is accelerating upward; that makes it look like the light beam moves on a curved trajectory." And Einstein says: "They are the same thing!"

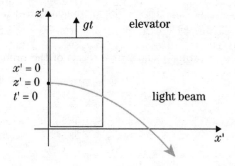

Figure 6: Trajectory of a light beam in the *elevator* reference frame.

This proved to him that a gravitational field must bend a light ray. As far as I know, no other physicist understood this at the time.

In conclusion, we have learned that it is useful to think about curvilinear coordinate transformations in space-time.

When we do think about curvilinear coordinates transformations, *the form of Newton's laws changes*. One of the things that happen is that apparent gravitational fields materialize, which are physically indistinguishable from ordinary gravitational fields.

Well, are they really physically indistinguishable? For some purposes yes, but not for all. So let's turn now to real gravitational fields, namely gravitational fields of gravitating objects like the Sun or the Earth.

Tidal Forces

Figure 7 represents the Earth, or the Sun, or any massive body. The gravitational acceleration doesn't point vertically on the page. It points toward the center of the body.

It is pretty obvious that there is no way that you could do a coordinate transformation like we did in the preceding section that would remove the effect of the gravitational field. Yet, if you are in a small laboratory in space and that laboratory is allowed to simply fall toward Earth, or toward whatever massive object you are considering, then you will think that in that laboratory there is no gravitational field.

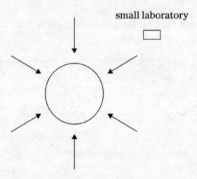

Figure 7: Gravitational field of a massive object, and small laboratory falling toward the object, experiencing inside itself no gravitation.

Exercise 1: If we are falling freely in a uniform gravitational field, prove that we feel no gravity and that things float around us like in the International Space Station.

But, again, there is no way *globally* to introduce a coordinate transformation that is going to get rid of the fact that there is a gravitational field pointing toward the center. For instance, a very simple transformation similar to equations (12) might get rid of the gravity in a small portion on one side of the Earth, but the same transformation will increase the gravitational field on the other side. Even more complex transformations would not solve the problem.

One way to understand why we can't get rid of gravity is to think of an object that is not small compared to the gravitational field. My favorite example is a 2000-mile man who is falling in the Earth's gravitational field; see figure 8. Because he is so big, different parts of his body feel different gravitational fields. Remember that the farther away you are, the weaker is the gravitational field.

His head feels a weaker gravity than his feet. His feet are being pulled harder than his head. He feels like he is being stretched, and that stretching sensation tells him that there is a gravitating object nearby. The sense of discomfort that he feels, due to the nonuniform gravitational field, cannot be removed by switching to a free-falling reference frame. Indeed, no change of mathematical description whatsoever can change this physical phenomenon.

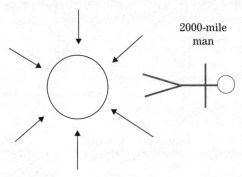

Figure 8: A 2000-mile man falling toward Earth.

The forces he feels are called *tidal forces*, because they play an important role in the phenomenon of tides, too. They cannot be removed by a coordinate transformation. Let's also see what happens if he is falling not vertically but sideways, staying perpendicular to a radius. In that case his head and his feet will be at the same distance from Earth. Both will be subjected to the same force in magnitude pointing to Earth. But since the force directions are radial, they are not parallel. The force on his head and the force on his feet will both have a component along his body. A moment's thought will convince us that the tidal forces will compress him, his feet and head being pushed toward each other. This sense of compression is again not something that we can remove by a coordinate transformation. Being stretched or shrunk, or both, by the Earth's gravitational field – if you are big enough – is an invariant fact.

In summary, it is not quite true that gravity is equivalent to going to an accelerated reference frame.

Andy: *Aha! So Einstein was wrong after all.*

Lenny: *Well, Einstein was wrong at times, but no, Andy, this was not one of those times. He just had to qualify his statement and make it a bit more precise.*

What Einstein really meant was that small objects, for a small length of time, cannot tell the difference between a gravitational field and an accelerated frame of reference.

It raises the following question: if I present you with a force field, does there exist a coordinate transformation that will make it vanish? For example, the force field inside the elevator, associated with its uniform acceleration with respect to an inertial reference frame, was just a vertical force field pointing downward and uniform everywhere. There was a transformation canceling it: simply use z- instead of z'-coordinates. It is a nonlinear coordinate transformation. Nevertheless, it gets rid of the force field.

With other kinds of coordinate transformations, you can make the gravitational field look more complicated, for example transformations that affect also the x-coordinate. They can make the gravitational field bend toward the x-axis. You might simultaneously accelerate along the z-axis while oscillating back and forth on the x-axis. What kind of gravitational field do you see? A very complicated one: it has a vertical component and it has a time-dependent oscillating component along the x-axis.

If instead of the elevator you use a merry-go-round, that is, a carousel, and instead of the (x', z', t) coordinates of the elevator, you use polar coordinates (r, θ, t), an object that in the stationary frame was fixed, or had a simple motion like the light beam, may have a weird motion in the frame moving with the merry-go-round. You may think that you have discovered some repulsive gravitational field phenomenon. But no matter what, the reverse coordinate change will reveal that your apparently messy field is only the consequence of a coordinate change. By choosing funny coordinate transformations, you can create some pretty complicated fictitious, apparent, also called *effective*, gravitational fields. Nonetheless they are not genuine, in the sense that they don't result from the presence of massive objects.

If I give you the field everywhere, how do you determine whether
it is fictitious or genuine, i.e., whether it is just the sort of fake
gravitational field resulting from a coordinate transformation to
a frame with all kinds of accelerations with respect to a simple
inertial one, or it is a real gravitational field?

If we are talking about Newtonian gravity, there is an easy way.
You just calculate the tidal forces. You determine whether that
gravitational field will have an effect on an object that will cause
it to squeeze and stretch. If calculations are not practical, you
take an object, a mass, a crystal. You let it fall freely and see
whether there were stresses and strains on it. If the crystal is big
enough, these will be detectable phenomena. If such stresses and
strains are detected, then it is a real gravitational field as opposed
to only a fictitious one.

On the other hand, if you discover that the gravitational field
has no such effect, that any object, wherever it is located and let
freely to move, experiences no tidal force – in other words, that
the field has no tendency to distort a free-falling system – then
it is a field that can be eliminated by a coordinate transformation.

Einstein asked himself the question: what kind of mathematics
goes into trying to answer the question of whether a field is a
genuine gravitational one or not?

Non-Euclidean Geometry

After his work on special relativity, and after learning of the mathematical structure in which Minkowski[7] had recast it, Einstein
knew that special relativity had a geometry associated with it. So
let's take a brief rest from gravity to remind ourselves of this important idea in special relativity. Special relativity was the main
subject of the third volume of TTM. Here, however, the only thing
we are going to use about special relativity is that space-time has
a geometry.

---

[7]Hermann Minkowski (1864–1909), Polish-German mathematician and
theoretical physicist.

In the Minkowski geometry of special relativity, there exists a kind of distance between two points, that is, between two events in space-time; see figure 9.

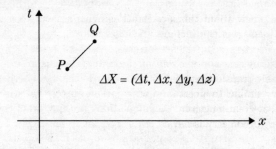

Figure 9: Minkowski geometry: a 4-vector going from P to Q.

The distance between P and Q is not the usual Euclidean distance that we could be tempted to think of. It is defined as follows. Let's call ΔX the 4-vector going from P to Q. To the pair of points P and Q we assign a quantity denoted $\Delta \tau$, defined by

$$\Delta \tau^2 = \Delta t^2 - \Delta \dot{x}^2 - \Delta y^2 - \Delta z^2$$

Notice that $\Delta \tau$ does not satisfy the usual properties of a distance. In particular, $\Delta \tau^2$ can be positive or negative; and it can be zero for two events that are not identical. The reader is referred to volume 3 of TTM for details. Here we only give a brief refresher.

The quantity $\Delta \tau$ is called the *proper time* between P and Q. It is an invariant under Lorentz transformations. That is why it qualifies as a sort of distance, just as in three-dimensional (3D) Euclidean space the distance between two points, $\Delta x^2 + \Delta y^2 + \Delta z^2$, is invariant under isometries.

We also define a quantity Δs by

$$\Delta s^2 = -\Delta t^2 + \Delta x^2 + \Delta y^2 + \Delta z^2$$

We call Δs the *proper distance* between P and Q. Of course, $\Delta \tau$ and Δs are not two different concepts. They are the same – just differing by an imaginary factor i. They are just two ways to talk about the Minkowski "distance" between P and Q. Depending on which physicist is writing the equations, they will rather use $\Delta\tau$ or Δs as the distance between P and Q.

Einstein knew about this non-Euclidean geometry of special relativity. In his work to include gravity, and to investigate the consequences of the equivalence principle, he also realized that the question we asked at the end of the previous section – are there coordinate transformations that can remove the effect of forces? – was very similar to a certain mathematics problem that had been studied at great length by Riemann. It is the question of deciding whether a geometry is flat or not.

Riemannian Geometry

What is a flat geometry? Intuitively, it is the following idea: the geometry of a page is flat. The geometry of the surface of a sphere or a section of a sphere is not flat. The *intrinsic geometry* of the page remains flat even if we furl the page like in figure 10. We will expound mathematically on the idea in a moment.

page flat   **page furled**

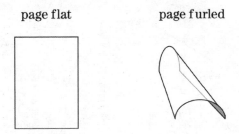

Figure 10: The intrinsic geometry of a page remains flat.

For now, let's just say that the intrinsic geometry of a surface is the geometry that a two-dimensional bug roaming on it, equipped with tiny surveying tools, would see if it were trying to establish an ordnance survey map of the surface.

If the bug worked carefully, it might see hills and valleys, bumps and troughs, if there were any, but it would not notice that the

page is furled. We see it because *for us* the page is embedded in
the 3D Euclidean space we live in. By unfurling the page, we can
make its flatness obvious again.

Einstein realized that there was a great deal of similarity in the two
questions of whether a geometry is non-flat and whether a space-
time has a real gravitational field in it. Riemann had studied the
first question. But Riemann had never dreamt about geometries
that have a minus sign in the definition of the square of the distance. He was thinking about geometries that were non-Euclidean
but were similar to Euclidean geometry – not Minkowski geometry.

Let's start with the mathematics of Riemannian geometry, that
is, of spaces where the distance between two points may not be
the Euclidean distance, but in which the square of the distance is
always positive.[8]

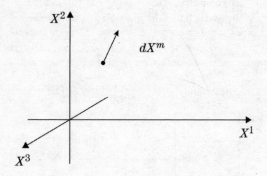

Figure 11: Small displacement between two points in a space.

We look at two points in a space; see figure 11. In our example
there are three dimensions, therefore three axes, X^1, X^2, and X^3.
There could be more. Thus a point has three coordinates, which
we can write as X^m, where m is understood to run from 1 to 3
or to whatever number of axes there is. And a little shift between
one point and another nearby has three components, which can
be denoted ΔX^m or, if it is to become an infinitesimal, dX^m.

---

[8]In mathematics, they are called *positive definite distances*.

If this space has the usual Euclidean geometry, the square of the length of dX^m is given by Pythagoras theorem

$$dS^2 = (dX^1)^2 + (dX^2)^2 + (dX^3)^2 + \dots \qquad (14)$$

If we are in three dimensions, then there are three terms in the sum. If we are in two dimensions, there are two terms. If the space is 26-dimensional, there are 26 of them and so forth. That is the formula for Euclidean distance between two points in Euclidean space.

For simplicity and ease of visualization, let's focus on a two-dimensional space. It can be the ordinary plane, or it can be a two-dimensional surface that we may visualize embedded in 3D Euclidean space, as in figure 12.

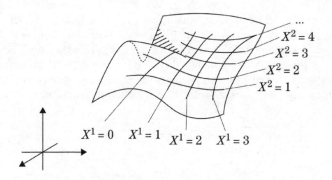

Figure 12: Two-dimensional manifold (i.e., 2D surface) and its curvilinear coordinates viewed embedded in ordinary 3D euclidean space.

There is nothing special about two dimensions for such a surface, except that it is easy to visualize. Mathematicians think of "surfaces" even when they have more dimensions. Usually they don't call them surfaces but *manifolds* or sometimes *varieties*.

Gauss had already understood that on curved surfaces the formula for the distance between two points was more complicated in general than equation (14). Indeed, we must not be confused by the fact that in figure 12 the surface is shown embedded in the usual three-dimensional Euclidean space. This is just for convenience of representation. We should think of the surface as a space in itself, equipped with a coordinate system to locate any point, with curvy lines corresponding to one coordinate being constant, etc., and where a distance has been defined. We must forget about the embedding 3D Euclidean space. The distance between two points on the surface is certainly not their distance in the embedding 3D Euclidean space, and is not even necessarily defined on the surface with the equivalent of equation (14). We will come shortly to exactly how these distances are represented mathematically.

Riemann generalized these surfaces and their metric (the way to compute distances) to any dimensions. But let's continue to use our picture with two dimensions in order to sustain intuition. And let's go slowly, so as not to miss any important detail.

The first thing we do with a surface is put some coordinates on it, which will allow us to quantify various statements involving its points. We just lay out coordinates as if drawing them with a piece of chalk. We don't worry at all about whether the coordinate axes are straight lines or not, because for all we know when the surface is a really curved surface, there probably won't even be things that we can call straight lines. We still call them X's.

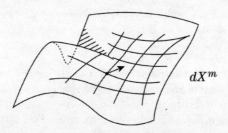

Figure 13: Two neighboring points and the shift dX^m between them.

The values of the X's are not related directly to distances. They are just numerical labels. The points $(X^1 = 0,\ X^2 = 0)$ and $(X^1 = 1,\ X^2 = 0)$ are not necessarily separated by a distance of one. Now we take two neighboring points; see figure 13. The two neighboring points are again related by a shift of coordinates.

But, unlike in figure 11, which was still a Euclidean space, now we are on an arbitrary curved surface with arbitrary coordinates.

Now we *define* a distance on the surface for points separated by a small shift like dX^m. It won't be as simple as equation (14), though it will have similarities with it. Here is the new definition of dS^2:

$$dS^2 = \sum_{m,\, n} g_{mn}(X)\ dX^m dX^n \tag{15}$$

The functions $g_{mn}(X)$'s, considered altogether, form what is called the *metric of the space*. It is a set of functions of the position X on the manifold under consideration.

Formula (15) is very general and applies whether the manifold is flat or curved. It is a very important formula in Riemannian geometry, and – we will soon see – even in the Minkowski geometry of relativity.

We will also see in a moment how the Einstein summation convention will enable us to rewrite formula (15) in a lighter form. The convention is explained in the section "Mathematical Interlude: Einstein Summation Convention," see infra.

Incidentally, formula (15) applies even to flat geometries equipped with curvilinear coordinates. Suppose that you take a flat geometry, like the surface of the page, but for some reason you use some curvilinear coordinates to locate points, and you ask what is the distance between two points close to each other. Then in general, the square of the distance between two points close to each other will be a quadratic form in the coordinate shifts dX^m's. A quadratic form means a sum of terms, each of which is the product of two little coordinate shifts, times a coefficient like g_{mn} that depends on X.

The surface of the Earth offers a simple example of distance on a curved manifold. Look at the distance between two nearby points characterized by longitude and latitude, as shown in figure 14. Let's denote with R the Earth radius. We take two points $(\phi,\ \theta)$ and $(\phi + d\phi,\ \theta + d\theta)$, where θ is the latitude and ϕ the longitude.

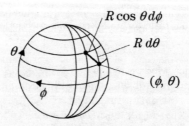

Figure 14: Formula for distance on the Earth surface. Shown are the points $(\phi,\ \theta)$ and $(\phi + d\phi,\ \theta + d\theta)$, and the segment joining them.

Apply Pythagoras theorem in a small, approximately flat, rectangular region to compute the square of the length of its diagonal. One pair of sides along a meridian have length $Rd\theta$. The other pair of sides along a parallel have length $Rd\phi$ but corrected by the cosine of the latitude. At the equator it is the full $Rd\phi$, whereas at the pole it is zero.

The general formula for the square of the distance is

$$dS^2 = R^2 \left[d\theta^2 + (\cos\theta)^2 d\phi^2 \right] \qquad (16)$$

It is an example of squared distance not just equal to $d\theta^2 + d\phi^2$ but having some coefficient functions in front of the differentials. In this case the interesting coefficient function is $(\cos\theta)^2$ in front of one of the terms $(dX^m)^2$. Note that the coefficient $(\cos\theta)^2$ is often written $\cos^2\theta$. Note also that in this case there are no terms of the form $d\theta\,d\phi$ because the natural curvilinear coordinates we chose on the sphere are still orthogonal at every point.[9]

In other examples – on the sphere with more involved coordinates, or on a more general curved surface like in figure 13 – where the coordinates are not necessarily locally perpendicular, the formula for dS^2 would be more complicated and comprise terms in $dX^m dX^n$. But it will still be a quadratic form. There will never be $d\theta^3$ terms. There will never be things linear. Every term will

---

[9]Note that spherical coordinates like we use here, which are a bit more sophisticated than Cartesian coordinates, were already much used in the sixteenth century, while Cartesian coordinates began to be used in analytic geometry only in the seventeenth century.

always be quadratic. Moreover in Riemannian geometry, at every point X, the quadratic form defining the metric locally is always positive definite.

You may wonder why we define the distance dS only for small (actually infinitesimal) displacements. The reason is that to talk about distance between two points A and B far away from each other, we must first of all define what we mean. There may be bumps and troughs in between. We could mean the shortest distance as follows: we put a peg at A and a peg at B and pull a string as tight as we can between the two points. That would define one notion of distance. Of course, there might be several paths with the same value. One might go around the hill this way. Then the other would go around the hill that way. Simply think on Earth of going from the North Pole to the South Pole.

Furthermore, even if there is only one answer, we have to know the geometry on the surface everywhere in the whole region where A and B are located, not only to calculate the distance but to know actually where to place the string. Therefore the notion of distance between any two points is more complicated than in Euclidean geometry. But between two *neighboring points* it is not so complicated. That is because locally a smooth surface can be approximated by the tangent plane and the curvilinear coordinate lines by straight lines – not necessarily perpendicular but straight.

Metric Tensor

Let's go deeper into the geometry of a curved surface and its links with equation (15), which defines the distance between two neighboring points on it. Recall the equation

$$dS^2 = \sum_{m,\ n} g_{mn}(X)\ dX^m dX^n$$

In order to get a feel about the geometry of the surface and its behavior, let's imagine that we arrange elements from a Tinkertoy Construction Set along the curved surface. For instance, they could approximately follow the coordinate lines on the surface. We would also add more rigid elements diagonally. This would create a lattice as shown in figure 15. But any reasonably dense lattice, sort of triangulating the surface, would do as well. Suppose furthermore that the Tinkertoy elements are hinged together in a way that lets them freely move in any direction from each other.

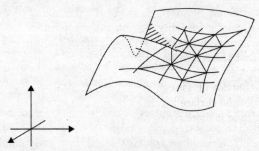

Figure 15: Lattice of rigid Tinkertoy elements arranged on the surface.

Imagine that we lift our lattice from the surface. Sometimes it will keep its shape rigidly, sometimes it won't. It will not keep its shape if it is possible to go from the initial shape to a new shape without forcing any Tinkertoy element to be stretched or compressed or bent.

In some cases it will even be possible *to lay it out flat*. It is the case, for instance, in figure 10 going from the shape on the right to the shape on the left – which is just a flat page.

> **Exercise 2**: Is it possible to find a curved surface and a lattice of rods arranged on it that cannot be flattened out, but can change shape?

Answer: Yes. According to Gauss's Theorema Egregium, which we invite the reader to look up, a surface can be modified without stretching or compressing it as long as we preserve everywhere its Gaussian curvature. For instance, it is possible to change in such a way a section of a hyperbolic paraboloid.

We shall see that the initial surface being able to take other shapes or not corresponds to the g_{mn}'s of equation (15) having certain mathematical properties.

The collection of g_{mn}'s has a name. It is called the *metric tensor*. It is the mathematical object that enables us to compute the distance between two neighboring points on our Riemannian surface.

Mind you, the g_{mn}'s are functions of X (the points of the manifold). So, strictly speaking, we are talking about a *tensor field*. But it is customary to talk casually of the metric tensor, keeping in mind that the collection of its components depends on X.

When the lattice of Tinkertoy elements can be laid out flat, the geometry of the surface is said to be *intrinsically flat*, or just *flat*. We will define it more rigorously later.

Sometimes, on the other hand, the lattice of little rods cannot be laid out flat. For example on the sphere, if we initially lay out a lattice triangulating a large chunk of the sphere, we won't be able to lay it out on a flat plane.[10]

The question we have to address is this: if I made a lattice of little rods covering a surface, and I gave you the length of each rod, without yourself building the lattice how could you tell me whether it is a flat space or an intrinsically curved space, which cannot be flattened and laid out on a flat plane?

Let's formulate the problem more precisely and mathematically. We start from the metric tensor $g_{mn}(X)$, which is a function of position, in some set of coordinates. Keep in mind that there are many different possible sets of curvilinear coordinates on the surface, and in every set of coordinates the metric tensor will look different. It will have different components, just like the same 3-vector in ordinary 3D Euclidean space has different components depending on the basis used to represent it, but in addition the components will vary with position in different ways.

I select one set of coordinates and I give you the metric tensor of my surface. In effect I tell you the distance between every pair of neighboring points. The question is: is my surface flat or not?

---

[10]This is a well-known problem of cartographers, which led to the invention of various kinds of maps of the world, the most famous being the Mercator projection map invented by Flemish cartographer Gerardus Mercator (1512–1594).

To answer that question, you may think of "checking Pi." Here is the way it would go. Think of a 2D surface embedded in the usual 3D Euclidean space as shown in figure 12. You select a point and mark out a disk around it. Then you measure its radius r as well as its circumference l, and you divide l by $2r$. If you get 3.14159... you would say that the surface is flat. Otherwise you would say that it is not flat, it is intrinsically curved. Notice that this procedure is good for a two-dimensional surface, under certain conditions. Anyway it is not so great for higher-dimensional surfaces.

What is the mathematics of taking a metric tensor and asking if its space is flat? What does it mean for it to be flat? By definition, it means this:

The space is flat if we can find a coordinate transformation, that is, a different set of coordinates, in which, at any point on the surface, *the distance formula for dS^2 becomes just $(dX^1)^2 + (dX^2)^2 + \ldots + (dX^n)^2$, as it would be in Euclidean geometry.*

It is not necessary that the *initial* $g_{mn}(X)$'s form everywhere the unit matrix, with ones on the diagonal and zeroes elsewhere – as if equation (15) were just Pythagoras theorem. But we must find a coordinate transformation that brings it to that form.

In that sense, it has a vague similarity with the question of whether you can find a coordinate transformation that removes the gravitational field. In fact, it turns out not to be a vague similarity at all but a close parallel. The question is: can we find a coordinate transformation that removes the curvy character of the metric tensor g_{mn}?

To answer that geometric question, we have to do some mathematics essential to relativity. It is not possible to understand general relativity without it. The mathematics is tensor analysis plus some differential geometry. At first it looks annoying because we have to deal with all these indices floating around, and different coordinate systems, and partial derivatives of components, etc. But once we get used to it, we will see that it is simple. It was invented, as said, by Ricci-Curbastro and Levi-Civita at the end of the nineteenth century to build on works of Gauss and Riemann. It was further simplified by Einstein, who set rules for the position of indices and astutely got rid of most summation symbols.

Before explaining what is the Einstein summation convention eliminating most summation symbols, let's spend a few moments explaining the simple concept of dummy variable.

Mathematical Interlude: Dummy Variables

We are accustomed to equations where all the variables have a substantial mathematical or physical meaning. A physical example is equation (7), reproduced here:

$$L(t) = \frac{1}{2}gt^2$$

This famous equation was found by Galileo Galilei in the first half of the seventeeth century,[11] before the invention of calculus. In fact, it is one of the equations that triggered the invention of calculus by Newton and Leibniz. It describes the fall of an object: L stands for the distance of fall as a function of time, g stands for the acceleration on the surface of the Earth, and t stands for time.

Another even simpler and purely mathematical example is

$$A = ab$$

where a is the length of a rectangle, b is its width, and A is its area.

But we are also familiar with equations where one of the variables is only a handy mathematical notation without a substantial meaning. A simple example is the well-known identity expressing the value of the sum of all the squares of the integers from 1 to m

$$\frac{m(m+1)(2m+1)}{6} = \sum_{n=1}^{n=m} n^2$$

---

[11]We give it here in its modern form. Galileo (1564–1642) just wrote that the distance of the fall was proportional to the square of the fall duration, which, if you think of it, is a mind-blowing discovery.

Here the variable m has a substantial meaning: it is the number up to which we sum. But the variable n on the right-hand side does not have such a substantial meaning. We could rewrite the equation as

$$\frac{m(m+1)(2m+1)}{6} = \sum_{k=1}^{k=m} k^2$$

It would be exactly the same equation.

The variable n, or the variable k, is called a *dummy variable*. It is only used to conveniently express the sum.

We will meet many formulas containing one or several dummy variables, usually expressing sums, in general relativity. They are so frequent that Einstein came up with a rule to simplify them. His rule, or convention, turned out to be not only a great simplification, but also a very useful notational device to write general relativity equations, providing a guide rail as well as having a meaning on its own. The convention is the topic of the next mathematical interlude. Later in this lecture and in the rest of the book, we will discover its remarkable usefulness.

Mathematical Interlude: Einstein Summation Convention

As we go along, we will see that certain patterns keep recurring in the equations. One such pattern involves expressions in which an index such as μ is repeated in a single expression. Here is an example. For the moment it doesn't matter what it means; it's just a pattern that we will see over and over.

$$\sum_{\mu} V^{\mu} U_{\mu}$$

There are a few things to note. First of all, there is a summation over μ, which means that μ is a *dummy index*. It is just another name, in the specific context of vectors and tensors, for a dummy variable. As a consequence, what letter we use doesn't matter. The expressions with μ, as above, or with ν, as below, represent exactly the same thing, whence, as we saw, the term *dummy*.

$$\sum_{\nu} V^{\nu} U_{\nu}$$

Secondly, the dummy index appears twice in the same expression
– not once, not three times, twice.

Finally, the repeated index occurs once as a *superscript* and once
as a *subscript*. I often say that it appears once upstairs and once
downstairs. That's the pattern: a sum over an index that appears
once upstairs and once downstairs.

Einstein's famous trick – the so-called *Einstein summation convention* – was just to leave out the summation sign. The rule is:
whenever we see something like $V^\mu U_\mu$, we automatically sum over
the index μ.

We can readily apply the convention to formula (15) that we met
earlier expressing the general form of the metric in a Riemannian
space (or for that matter in a Minkowskian space as well, we shall
see). It was

$$dS^2 = \sum_{m,\,n} g_{mn}(X)\, dX^m dX^n$$

With the Einstein summation convention it becomes

$$dS^2 = g_{mn}(X)\, dX^m dX^n$$

Simpler! Isn't it?

Usually, not forgetting that the g_{mn}'s components depend on X,
i.e., remembering that the metric tensor is actually a tensor field,
we simplify it even further to

$$dS^2 = g_{mn}\, dX^m dX^n$$

Andy: *Did it really take Einstein to invent the summation convention?*

Lenny: *I guess it did. When I was a student, I read Einstein's
famous 1916 paper "The Foundation of the General Theory of
Relativity." It was my habit when I learned new physics to write
out the equations as I read them. At the start of the paper, the
equations were written as anyone else would write them. Here's
his equation 2:*

$$dX_\nu = \sum_\sigma a_{\nu\sigma} dx_\sigma$$

*But then all of a sudden, right after equation 7, Einstein casually
remarks that there is always a summation when indices appear
twice.*[12] *So from now on, he said, we'll just keep that in mind and
stop writing the summation sign. It's pretty clear that he just got
tired of writing them. I was pretty tired of writing them too. What
a relief it was.*

End of interlude on Einstein summation convention.

Let's return to the metric and its various forms in several different coordinate systems. To find a set of coordinates that make
equation (15) become equation (14) is a more involved procedure
than just diagonalizing the matrix g_{mn}. The reason is that there
is not one matrix. As we stressed, each component g_{mn} depends
on X.

It is the same tensor field, but it has a different matrix at each
point.[13] You cannot diagonalize them all at the same time. At a
given point, you can indeed diagonalize $g_{mn}(X)$ even if the surface is not flat. It is equivalent to working locally in the tangent
plane of the surface at X, and orthogonalizing the coordinate axes
there. But you cannot say that a surface is flat because it can be
made at any given point locally to look like the Euclidean plane.

Let's examine equation (14) more closely. It can be written in
terms of a special matrix whose components are the *Kronecker-
delta symbol* δ_{mn}, defined in the following way.[14]

First of all, δ_{mn} is zero unless $m = n$. For example, in three di-
mensions δ_{12}, δ_{13}, and δ_{23} are all zero, but δ_{11}, δ_{22}, and δ_{33} are
nonzero. In other words, at each point the Kronecker-delta sym-
bol is a diagonal matrix.

---

[12]Later, Einstein devised the superscript and subscript notations for the
indices of tensors, and his rule henceforth applied only to pairs of indices,
with the same dummy variable, one upstairs and the other downstairs.

[13]For a given set of coordinates, we have a collection of matrices – one at
each point. For another set of coordinates, we will have another collection
of matrices. At each point, the *components* of the tensor depend on the
coordinates, but the tensor itself is an abstract object that doesn't. We
already met the distinction with 3D vectors.

[14]Named after the German mathematician Leopold Kronecker
(1823–1891).

Secondly, the diagonal elements are all equal to 1:

$$\delta_{11} = \delta_{22} = \delta_{33} = 1$$

Armed with the Kronecker-delta and the Einstein summation convention, we can rewrite equation (14) in the compact form,

$$dS^2 = \delta_{mn}\, dX^m dX^n \tag{17}$$

To determine if a space is flat, we look for a coordinate transformation, $X \to Y$, that turns g_{mn} into δ_{mn} everywhere. Remember that X and Y represent the *same point* P. This point P is simply located with two different reference systems, which, as we stressed, are nothing more than some geometric labeling procedure.

Later, the points P will be events in space-time, and the Kronecker-delta will be replaced by a slightly more involved diagonal matrix in Minkowski geometry (also called Minkowskian or Einsteinian geometry), but many of the ideas will remain unchanged. However let's not go too fast, and for the moment let's stay in Riemannian geometry. Riemannian geometry is everywhere locally Euclidean. It can be thought of as "Euclidean geometry on a piece of rubber."

For most metrics it is not possible to find a coordinate transformation that transforms everywhere the g_{mn} into δ_{mn}. It is only when the space is intrinsically flat that we can.

In summary, I give you the metric tensor of my surface, that is, the g_{mn} of equation (15), which we now write

$$dS^2 = g_{mn}(X)\, dX^m dX^n$$

The question I ask you is: can you, by a coordinate transformation $X \to Y$, reduce it to equation (17)? That is, in the Y system,

$$dS^2 = \delta_{mn}\, dY^m dY^n$$

There is no need to write $\delta_{mn}(Y)$, since the Kronecker-delta symbol by definition has a unique form. However, for the sake of clarity, we will sometimes still write $\delta_{mn}(Y)$ because it reminds us of which system of coordinates we are using.

If the answer is yes, the space is called *flat*. If it is no, the space is called *curved*. Of course, the space could have some portions that are flat. There could exist a set of coordinates such that in a region the metric tensor is the Kronecker-delta. But the surface is called flat only if it is everywhere flat.

This becomes a pure mathematics problem: given a tensor field $g_{mn}(X)$ on a multidimensional space (which mathematicians call a *manifold*), how do we figure out if there is a coordinate transformation that would change it into the Kronecker-delta symbol?

To answer that question, we have to understand better how things transform when we make coordinate transformations. That is the subject of *tensor analysis*. We begin to present the subject in the rest of this lecture, and will treat it in more depth in lecture 2.

The analogy between tidal forces and curvature actually is not an analogy, it is a very precise equivalence. In the general theory of relativity, the way you diagnose tidal forces (or said more accurately, their generalization) is by calculating the curvature tensor. A flat space is defined as a space where the curvature tensor is zero everywhere. Therefore it is a very precise correspondence. Simply stated:

Gravity is curvature.

But we will come to this conclusion as we get through tensor analysis. Obviously, in trying to determine whether we can transform away $g_{mn}(X)$ and turn it into the trivial $\delta_{mn}(Y)$, the first question to ask is: how does $g_{mn}(X)$ transform when we change coordinates? We have to introduce notions of tensor analysis that are rather easy.

We shall express the first tensor rule, then present a mathematical interlude spelling out some general facts on vectors and tensors, then present the second tensor rule.

We will conclude this copious lecture again with some general considerations on covariant and contravariant components of vectors and tensors.

First Tensor Rule: Contravariant Components of Vectors

Sometimes tensor notations are a bit of a nuisance because of all the indices. At first we can get confused by them. But soon we will discover that the manipulations obey strict rules and turn out to be rather simple.

We shall begin with a simpler thing than $g_{mn}(X)$. Suppose that there are two sets of coordinates on our surface: a set of coordinates X^m, and a second set that we could call X' as we did earlier. But then we would be running into horrible notations with cluttered expressions like X'^1. So we denote the second set of coordinates Y^m. To be very explicit, if we are on a space of dimension N, the same point P has coordinates

$$\left[\ X^1(P),\ X^2(P),\ \ldots\ , X^N(P)\ \right]$$

and also has coordinates

$$\left[\ Y^1(P),\ Y^2(P),\ \ldots\ , Y^N(P)\ \right]$$

The X's and Y's are related because if you know the coordinates of a point P in one set of coordinates, then in principle you know where the point is. Therefore you also know its coordinates in the other coordinate system. Thus each coordinate X^m is a function of all the coordinates Y^n. We can use whatever dummy index we want if that helps avoid confusion. We will simply write

$$X^m(Y)$$

Likewise each Y^m is assumed to be a known function of all the X^n's:

$$Y^m(X)$$

In short, we have two coordinate systems, each one a function of the other. The correspondence is one-to-one since these are coordinate systems. And we assume that the functions are nice and smooth.

Now we ask: how do the differential elements dX^m transform? The collection of differential elements dX^m is a small vector, as shown in figure 16. Remember that the vector itself is a pair of points (an origin and an end). It is independent of the coordinate system. But in order to work with it, it is useful to express it using its components dX^m.

Figure 16: Small displacement expressed in the X coordinate system.

The notation dX^m is used to represent the small vector

$$dX^m = \left[\, dX^1,\ dX^2,\ \ldots\ ,dX^N \,\right]$$

Said another way, when we change X a little bit, the point P moves to a nearby point Q, and the displacement is dX^m.

Let's look at the same displacement, expressed in the Y coordinate system. We want to know how dY^m can be expressed in terms of the dX^p's. It is an elementary result of calculus that

$$dY^m = \sum_p \frac{\partial Y^m}{\partial X^p}\, dX^p$$

or using the summation convention,

$$dY^m = \frac{\partial Y^m}{\partial X^p}\, dX^p \tag{18}$$

Let's spell out even more explicitly what equation (18) says: the total change of some particular component Y^m is the sum of the rate of change of Y^m when you change only X^1, times the little change in X^1, namely dX^1, plus the rate of change of Y^m when you change only X^2, times the little change in X^2, namely dX^2, and so forth up to X^N and dX^N because equation (18) means a sum over the dummy index p going from 1 to N.

We now turn to some general considerations on vectors and tensors.

So far we have used several times the term *tensor* (tensor calculus, metric tensor, curvature tensor, first tensor rule, etc.), without explaining what is a tensor! As the reader has understood, it is a fundamental mathematical tool in general relativity. You may even remember that "it extends the concept of vector." But that is certainly not a sufficient explanation to grasp what it is.

We won't go into a full fledged exposition of linear algebra and tensors – which the reader may find in any good manual on the subject. However, as I have done several times in The Theoretical Minimum series, for instance, when I dared to explain in volume 1 integral calculus or partial differentiation in brief interludes of a few pages, because we needed those tools for classical mechanics, it is time in this lecture for a third mathematical interlude presenting in some detail vectors and tensors.

Mathematical Interlude:
Vectors and Tensors

Let's begin with the simplest notion of a tensor, namely a scalar. A scalar $S(X)$ is a function of position with the property that it has the same value in every coordinate system. For that reason, we could also denote it $S(P)$, but we want to insist on the coordinate system we chose to use, so we write instead $S(X)$. For the same scalar in the Y coordinate system, we will temporarily use the notation $S'(Y)$. (Later we will use $S(Y)$ and $S(X)$ for both, because it is clearer when we talk about the chain rule.)

Its transformation properties are trivial: it doesn't transform at all. An example drawn from meteorology would be the temperature at a point in space. The transformation property of a scalar reflects this triviality,

$$S'(Y) = S(X)$$

In the case of temperature, this says that the temperature at a point is just a number.[15] It does not depend on the orientation

---

[15] *Number* and *scalar* are two equivalent terms for the same thing. What is the reason for talking about "scalars"? Numbers are often called scalars because one number can always be obtained from another number with a change of scale: for instance, you can obtain 7 from 2, just by multiplying 2 by 3.5. You cannot do that with any pair of vectors. It is possible only with colinear vectors. Strictly speaking, the term *scalar* is reserved for real numbers. But we often also casually call complex numbers scalars.

of the coordinate system at that point. Note too that scalars do not have components, or perhaps more accurately, they have only one component: the value of the scalar itself.

Let's turn to the next simple kind of tensors, namely vectors. We shall see that there are two kinds.

We all have an intuitive idea of what a vector in a Riemannian geometry is. It is a little arrow, usually attached to a point in space. It points in a direction and it has a magnitude. An example, again from meteorology, would be the wind velocity.

In a Riemannian geometry, a vector is a thing unto itself, but given a coordinate system and a metric, it can be described by components in one of two ways: either contravariant or covariant components.

Since the terms can be a little confusing, let's stress right away that what are called the *contravariant components* of a vector are the good old components with which we construct the vector as a linear combination of the basis vectors.

We will see that we can also attach to a vector another set of numbers, called its *covariant components*. They are not its ordinary contravariant components, but something else, the geometrical meaning of which will be explained in lecture 2. The contravariant and covariant components of a vector will be simply related to each other with the help of the metric.

These components, like the components of the metric itself, will vary when the coordinates system changes. For the moment, however, let's not think of a metric, only of a system of coordinates X and a system of coordinates Y. We position ourselves at a point P. At this point, we consider a set of numbers attached to it and that depends on the coordinate system.

Disregarding any geometric interpretation, this set of numbers can be viewed as an abstract "vector." As said, we are in the case where the vector will change with the coordinate system.

In that case we will have two kinds of vectors: covariant or contravariant vectors. Notice I said covariant or contravariant *vectors* – not covariant or contravariant *components*. Later, when we have introduced a metric, we can put the two together to describe a single kind of vector (the intuitive arrow) in two ways.

What is it that makes a collection of numbers like dX^m a contravariant vector, rather than just a collection of numbers? The answer is the transformation properties under a coordinate transformation. Equation (18) defines the paradigm for the transformation of a contravariant vector.

A contravariant vector is a set of numbers V^m that transform as follows:

$$(V')^m = \frac{\partial Y^m}{\partial X^p} V^p \tag{19}$$

In this equation the variables V are the components of the vector in the X coordinate system and (V') are the components in the Y system. Looking back at equation (18), we see that the differential displacement dX^m is a contravariant vector.

There are a couple of things to note. First of all, I have used the summation convention so that the index p is summed over. Secondly, the index p in the expression $\partial Y^m/\partial X^p$ is a downstairs index. That's a convention that we have already mentioned in the interlude on Einstein summation convention and that the reader will have to remember: when an upstairs index occurs in the denominator of an expression, it counts as a downstairs index.

Generally speaking, in a "level" expression (i.e., with no denominator) or in the numerator of a fraction, a superscript index is called a *contravariant index*. And a subscript index is a called a *covariant index*. But, as we said, according to the summation convention, a superscript in the denominator of a fraction acts like a covariant index.

Let's move on to the second kind of vector – a covariant vector. If the iconic contravariant vector is the displacement dX^m, the iconic covariant vector is the gradient of a scalar $S(X)$.

Its components are given by the derivatives of the scalar along the coordinate axes:

$$\frac{\partial S(X)}{\partial X^p} \tag{20}$$

Clearly these components depend on the choice of coordinates, and will transform when the coordinates are transformed. For example, suppose we transform from the X to the Y system. To compute the components of the gradient in the Y system, we use a version of the chain rule of calculus (see lecture 2 of volume 1 of TTM, in which the chain rule is explained). We get

$$\frac{\partial S}{\partial Y^m} = \frac{\partial S}{\partial X^p} \frac{\partial X^p}{\partial Y^m} \tag{21}$$

From this we can abstract the general rule for the transformation of a covariant vector:

$$(W')_m = W_p \frac{\partial X^p}{\partial Y_m} \tag{22}$$

Thus, in equation (18), we met the *first example of transformation of a tensor*, because an ordinary vector, corresponding for instance to the position of a point, or to a displacement (in other words, a translation), or to a velocity, etc., is a contravariant vector, which is a simple kind of tensor.

Indeed, we now have the expressions, in two different coordinate systems, of the small displacement of a point on the surface (figure 16). They are dX^m and dY^m. Let's repeat that the dX^m and dY^m are two sets of components for the *same* displacement. And we know how to go from one set to the other.

Figure 17, which completes figure 16, shows the small displacement, and also locally the two sets of coordinates.

By now the reader has understood that equation (18) is simply the transformation property of the components of the displacement vector when this displacement vector (which is itself a well-defined geometric object, being defined independently of any coordinate system[16]) is expressed in the X system and in the Y system.

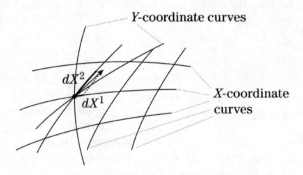

Figure 17: Small displacement, and two sets of coordinates. The small vector has components (dX^1, dX^2) shown, but also (dY^1, dY^2) not shown.

Note on terminology: because we will deal with vectors that can have contravariant expressions but also covariant expressions, we will prefer to speak of the contravariant components of a vector or the covariant components of a vector.

In short, the term *contravariant* comes from the fact that if we change the unit vectors in the coordinate system, for instance if we simply *divide* the length of each of them by ten, the components of a vector representing a translation will be *multiplied* by ten. Turning to the other term, *covariant* comes from the fact that, in the same kind of change of coordinates, the components of a gradient will be *divided* by ten.

---

[16]Notice, however, that it is difficult to speak of geometric concepts without some kind of coordinate system. The two famous American geometers Oswald Veblen (1880–1960) and John Whitehead (1904–1960), aware of the difficulty to define what is geometry, wrote in their book *The Foundations of Differential Geometry* that geometry is what experts call geometry. :-) This statement was considered outrageous by the Russian mathematician Andreï Kolmogorov (1903–1987) and his coauthors Alexander Alexandrov (1912–1999) and Mikhaïl Lavrentiev (1900–1980) in their famous book on mathematics, the English translation of which is *Mathematics: Its Content, Methods and Meaning*, MIT Press, 1969.

The interlude presented the simplest kind of tensors: tensors of rank 0, which are simply scalars; and tensors of rank 1, which are contravariant vectors and covariant vectors. The next kinds of tensors, of rank 2 or more, will be presented in the last section of this lecture.

Second Tensor Rule: Covariant Components of Vectors

Although we have already mentioned it cursorily in the preceding mathematical interlude, for the sake of symmetry, let's spell out the second tensor rule concerning the covariant components of vectors. These vectors are used to represent other things than position or translation or velocity or acceleration, etc. The reader may primarily think of gradients of scalar fields.

Examples of scalar fields are the temperature, the atmospheric pressure, the Higgs field, whatever has, at any point in the space, a value that is not multidimensional but simply a number, and that doesn't change if we change coordinates.

The wind velocity is not a scalar field because at every point it has a vector value. It is a vector field. It is important to note the following point, which should clarify things:

If we tried to consider only the first component of the vector representing the wind, we would not get a scalar field, because it would not be invariant under change of coordinates.

Thus the gradient of a scalar function is a vector (in the sense of a collection of components). But it is not an ordinary vector. Indeed, its components don't transform in the same way as do the contravariant components of ordinary vectors.

We saw earlier that an application of the chain rule gave us equation (21), which we reproduce here:

$$\frac{\partial S}{\partial Y^m} = \frac{\partial S}{\partial X^p} \frac{\partial X^p}{\partial Y^m}$$

Denoting by (W') the gradient of S with respect to the Y's, and by W its gradient with respect to the X's, it can be rewritten as equation (22), which we also reproduce, attributing it a new number:

$$(W')_m = \frac{\partial X^p}{\partial Y^m}\, W_p \qquad (23)$$

Equation (23) doesn't apply only to gradients; it is the fundamental equation linking the primed and unprimed versions of the covariant components of a vector, that is, its components in the Y system and in the X system.

Notice that the indices m of W' and p of W are downstairs. The index p is a dummy index that is to be summed over as it also appears upstairs in ∂X^p. It is a nice example of the very useful Einstein summation convention and of its smooth workings.

Let's rewrite equations (19) and (23) next to each other, and relabel them:

Contravariant components

$$(V')^m = \frac{\partial Y^m}{\partial X^p}\, V^p \qquad (24a)$$

Covariant components

$$(W')_m = \frac{\partial X^p}{\partial Y^m}\, W_p \qquad (24b)$$

They look very much alike except that $\partial Y^m / \partial X^p$ appears in the first one, and the inverse, $\partial X^p / \partial Y^m$, in the second.

Let's recall one last time that displacements, or positions, or velocities, etc., are described with vectors having contravariant components. We saw that these change *contrary* to the basis change.

Gradients, on the other hand, are described with vectors the components of which change *like* the basis change. That is why their components are called covariant. But these vectors are different from the somewhat more intuitive contravariant vectors.

In mathematics, vectors with covariant components are sometimes viewed as vectors in the dual space of the primary vector space under consideration. They are then *dual vectors* like linear forms are. But we won't adopt this approach. For us vectors will be things that have a one-indexed collection of contravariant components and also of covariant components.

Equations (24a) and (24b) are fundamental equations for this course. The reader needs to understand them, become familiar and at ease with them, because they are absolutely central to the entire subject of general relativity. You need to know where the indices go for different kinds of objects, and how these objects transform. That is in some sense what general relativity is all about: the transformation properties of different kinds of objects.

Covariant and Contravariant Components of Vectors and Tensors

We have seen two ways to think about an *ordinary* vector. First of all, we can think of it like we have learned in high school: it is a displacement with a length and a direction, that is, *an arrow* in a space. This is geometrically well defined even before we consider any basis.

We can also think of it more abstractly as some object that has components. These components depend on the basis. If the components transform in a certain way when we change basis, namely according to equation (24a), then the object behaves exactly like our good old vectors. Therefore we can also equate the object to an ordinary vector. In tensor analysis we call them vectors whose components are contravariant.

Similarly, some other objects have components that transform according to equation (24b). They cannot be equated to our old ordinary vectors, but to other geometric things. We mentioned that mathematicians view them as dual vectors. We will just call this second type of object vectors whose components are covariant. In fact, we will see in lecture 2 that our abstract vectors have a contravariant version and a covariant version.

In tensor calculus, of which general relativity makes heavy use,[17] paradoxically for those people who have a geometric mind or intuition, it is often useful, at least at first, to forget about the geometric interpretation of the objects we manipulate, and to focus only on how collections of numbers attached to points in our space behave when we change systems coordinates.

A vector – be it with contravariant or covariant components – is a special case of a tensor. Following what we just said, we are not going to define tensors geometrically. For us, at first, tensors will be things that are defined by the way they transform. The way they transform means the way they change (or if you prefer, their components change) when we go from one set of coordinates to another. Later we will give a geometric interpretation of some tensors. We will also go deeper into contravariant and covariant components. We will see that an object with one index can have a contravariant version and a covariant version. All this will be developed in the next lecture. For the time being, let's continue to proceed step by step in our construction of the mathematical tools necessary for general relativity.

The next step, for us now, is to talk about tensors with more than one index.

The best way to approach tensors with several indices is to consider a special, very simple case to start with. Let's imagine the "product" of two vectors with contravariant components.[18] We consider the two vectors with contravariant components, V and U, and we consider the following product:

---

[17]Einstein developed his ideas in special relativity without using tensor calculus nor even Minkowski geometry, which Minkowski, who had been Einstein's teacher at Zurich, introduced only in 1908. Poincaré also did some preliminary work in this direction. At first Einstein did not think that this heavy mathematical recasting of the theory of relativity was useful. But he soon changed his mind. When general relativity was completed, in 1915, Einstein said it would not have been possible without abstract non-Euclidean geometry and tensor calculus. Hermann Minkowski (1864–1909) did not participate in the development of general relativity because he died in 1909. His good friend David Hilbert (1862–1943) however, did play a role in 1915; see lecture 9.

[18]It is not the dot product nor the cross product. It is going to be called the *outer product* or *tensor product*. Anyway, it is an operation that to two things associates a third thing.

$$V^m U^n$$

Without further ado, we will now always use the convention that contravariant components, or contravariant indices refering to these components, are noted upstairs.

The vectors V and U don't have to come from the same space. If the dimensionality of the space of V is M, and the dimensionality of the space of U is N, there are $M \times N$ such products. As usual, we use the notation $V^m U^n$ to denote one product as well as the collection of all of them – just like V^m denotes one component of the vector V, but is also a notation, showing explicitly the position of the index, and therefore the nature of the full vector V itself.

Let's define T^{mn} as

$$T^{mn} = V^m U^n \qquad (25)$$

Notice that it matters where and in which order we write the indices of T^{mn}, because, for instance, T^{mn} is not the same as T^{nm}. The reader is invited to explain why. Soon we will also see combinations of indices upstairs and downstairs.

Product T^{mn} is a special case of tensor of rank 2. Rank 2 means that the collection of component products has two indices. It runs over two ranges: m runs from 1 to M, and n runs from 1 to N. For example, if both V and U come from a four-dimensional space, there will be 16 components $V^m U^n$. In that case T^{mn}, as we saw, represents one component but also the entire collection of 16 components.

How does T^{mn} transform?

For example V^m and U^n could be the components of the vectors V and U in the unprimed frame of reference, the reference frame using the X coordinates. Since we know how the individual components transform, when we go to the Y coordinates, we can figure out how T transforms. Let's call $(T')^{mn}$ the mn-th component of the tensor in the primed frame:

$$(T')^{mn} = (V')^m (U')^n$$

Then using equation (24a) twice, this can be rewritten as

$$(T')^{mn} = \frac{\partial Y^m}{\partial X^p} \, V^p \, \frac{\partial Y^n}{\partial X^q} \, U^q$$

The four terms on the right-hand side are just four numbers, so we can change their order and write it

$$(T')^{mn} = \frac{\partial Y^m}{\partial X^p} \, \frac{\partial Y^n}{\partial X^q} \, V^p \, U^q$$

Finally, $V^p \, U^q$ is just T^{pq}. So the way T transforms is

$$(T')^{mn} = \frac{\partial Y^m}{\partial X^p} \, \frac{\partial Y^n}{\partial X^q} \, T^{pq} \qquad (26)$$

We found in the special case of a product of ordinary vectors how T transforms. Now this leads us to the following definition:

Anything that transforms according to equation (26) is called a tensor of rank 2 with two contravariant indices.

If there were more indices upstairs, the rule would be adapted in the obvious manner. A tensor of rank 3, all indices contravariant, would transform like this:

$$(T')^{lmn} = \frac{\partial Y^l}{\partial X^p} \, \frac{\partial Y^m}{\partial X^q} \, \frac{\partial Y^n}{\partial X^r} \, T^{pqr}$$

What kinds of things are tensors like that? Many things. Products of vectors are particular examples, but there are other things that are not products and still are tensors according to this definition.

We are going to see that the metric object g_{mn} is a tensor. But it is a tensor with covariant indices. So to finish this lecture, let's see how things with covariant indices transform. Equation (24b) shows how an object with only one covariant index transforms. It is a tensor of rank 1 of covariant type.

Let's begin again with the particular case of the product of two
covariant vectors W and Z, or to speak less casually, two vectors
with covariant components. Their product transforms as follows:

$$(W')_m (Z')_n = \frac{\partial X^p}{\partial Y^m} \, \frac{\partial X^q}{\partial Y^n} \, W_p Z_q$$

Here we have discovered a new transformation property of a thing
with two covariant indices, that is, two downstairs indices.

More generally let's consider an object that we will denote T_{mn}.
It is no longer simply a product of vectors but a different object.
However, the letter T signals that it is something that will still be
a tensor. It is a tensor with two lower indices, and it transforms
according to this equation:

$$T'_{mn} = \frac{\partial X^p}{\partial Y^m} \, \frac{\partial X^q}{\partial Y^n} \, T_{pq} \tag{27}$$

Again, anything that transforms according to equation (27) is
called a tensor of rank 2 with two covariant indices.

It is left to the reader to figure out how a tensor with one upper
index and one lower index must transform.

In the next lecture, we will also see how the metric object g of
equation (15) transforms. We will see that it is a tensor with two
covariant indices.

Then the question we will ask is: given that equation (27) is the
transformation property of g, can we or can we not find a coordinate transformation that will turn g_{mn} into δ_{mn}?

That is the mathematics question. It is a hard question in general.
But we will find the condition.

Lecture 2: Tensor Mathematics

Andy: *These components you call contrarian, they really deserve their name! They are difficult to understand.*

Lenny: *Not contrarian, contravariant! It's simple enough, if you express your height in feet, it's about six. If you change for the smaller inches, it becomes 72. It varies contrary to the unit used.*

Andy: *Well, maybe, but I think I prefer covariant components. I've always preferred empathic people.*

Lenny: *Yeah, but sometimes a bit of contradiction is fruitful. Anyway, we will use both types of components.*

Introduction

In this lecture we study vectors and tensors in a Riemannian geometry. For us, it is an intermediary step toward our study of vectors and tensors in a Minkowski geometry. Remember that a Riemannian geometry is a cousin of the Euclidean geometry that all of us studied in high school. The distance between two close points is always positive, as the natural concept of distance should be. But it doesn't have the nice and simple global properties – with parallel lines, translations of everything as a block, etc. – of Euclidean geometry. To use an informal image, a Riemannian geometry is the intrinsic geometry of a 2D piece of rubber. And it can be extended to any number of dimensions.

However in general relativity, we will need an even more baroque geometry, where two distinct points, called *events*, in space-time can be separated by a distance whose square can be positive, null, or even negative. Such a geometry is also sometimes called *Minkowskian* or *Einsteinian* and will be studied in later lectures.

So for the time being, we are in a space with a Riemannian geometry. The Holy Grail we are seeking is a method to distinguish really flat spaces from really non-flat ones.

A good notation, as we said, will carry us a long way. When it is well conceived, it automatically tells us what to do next. That means that we can do physics in a completely mindless way – at least until we stumble upon the next subtle point.

It is like Tinkertoys. It is pretty clear how to assemble the elements. A stick can only go into a piece with a hole. You can try putting a hole into a hole or forcing a stick into a stick. There is only one thing you can do. You can put the stick into the hole, and the other end of the stick can go into another hole. Then there are more sticks and more holes you can put them into.

The notation of general relativity is much like that. If you follow the rules, you almost can't make mistakes. But you have to learn the rules. They are the rules of tensor algebra and tensor analysis.

Flat Space

The goal we are aiming at is to understand enough about tensor algebra and analysis, and metrics, to be able to distinguish a truly flat geometry from a truly non-flat geometry. That seems awfully simple. Flat means like a plane. Non-flat means with bumps and lumps in it. You would think we could tell the difference very easily. Yet sometimes it is not so easy.

For example, as discussed in the last lecture, if I consider the page of a book, its natural configuration is flat. If I roll it or furl it, the page can look curved but it is not really, intrinsically, curved. It is exactly the same page. The relationship between the parts of the page, the distances between the letters, the angles, and so forth, don't change. At least the distances between the letters *measured along the page* don't change. So a folded page, if we don't stretch it, if we don't modify the relations between its parts, doesn't acquire a new geometry. In particular, it doesn't acquire curvature.

Technically, folding the page introduces only what is called an *extrinsic curvature*. Extrinsic curvature has to do with the way a space – in our case the page – *is embedded in a higher-dimensional space*. For instance, whatever I do with the page is embedded in the three-dimensional space of the room. When the page is laid out flat on the desk, it is embedded in the 3D embedding space in one way. When it is furled, as shown in figure 10 of lecture 1, it is embedded in the same 3D space another way.

The extrinsic curvature we perceive has to do with *how* the space of the page is embedded in the larger space. But it has nothing to do with its *intrinsic geometry*. If you like, you can think of the intrinsic geometry as the geometry of a little bug who roams over the surface. It cannot look out of the surface. It only looks around while crawling along the surface. It may have surveying instruments with which it can measure distances along the surface. It can draw a triangle, measure also the angles within the surface, and do all kinds of interesting geometric studies. But it never sees the surface as embedded in a larger space. Therefore the bug will never detect that the page might be embedded in different ways in a higher-dimensional space. It will never detect it if we create a furl like in figure 10 of lecture 1, or if we remove the furl and flatten the page out again. The bug just learns about the intrinsic geometry. The intrinsic geometry of the surface means the geometry that is independent of the way the surface is embedded in a larger space.

General relativity and Riemannian geometry, and a lot of other
geometries, are all about the intrinsic properties of the geometry of the space under consideration. It doesn't have to be two-
dimensional. It can have any number of dimensions.

Another way to think about the intrinsic geometry of a space is
like this. Imagine sprinkling a bunch of points on the page – or
on a three-dimensional space, but then we would have to fiddle
with it in four dimensions or more.... Then draw lines between
them so they triangulate the space. Finally state what the distance between every pair of neighboring points is. *Specifying those
distances specifies the geometry.*

Sometimes that geometry can be flattened out without changing the length of any of these little links. In the case of a two-
dimensional surface, it means laying it out flat on the desk without stretching it, tearing it, or creating any distortion. Any small
equilateral triangle has to remain an equilateral triangle. Every
small little square has to remain a square, and so on. The small
square cannot be bent into a parallelogram because its diagonals
must also keep their length. All angles must be preserved.

But if the surface is intrinsically non-flat, there will be parts of
the surface that cannot be flattened out. The other day on his
motorbike, Andy saw on the road a bulge, probably due to pine
roots, with lines drawn on it, and the warning "watch the bump"
painted on the pavement. The road menders must have taken a
course in general relativity! Such a bump cannot be flattened out
without stretching or compressing some distances.

A curved space is basically one that, viewed or embedded into a
larger one, cannot be flattened out without distorting it. It is an
intrinsic property of the space, not an extrinsic one.

Metric Tensor

We want to answer the following mathematical question: given a
space and its metric defined by the equation

$$dS^2 = g_{mn}(X) \, dX^m dX^n \qquad (1)$$

is it flat or not?

It is important to understand that the space may be intrinsically curved, like the road with a bump, or we *may think* that it is curved because equation (1) looks complicated, when actually it is intrinsically flat.

For instance, we can draw on a flat page curvilinear coordinates as shown in figure 1.

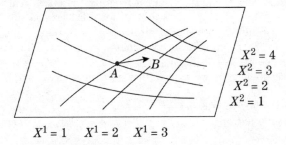

Figure 1: Curvilinear coordinates X's of a flat page.

Let's forget that we look comfortably at the page from our embedding 3D Euclidean space. At first sight the coordinate axes X's suggest that it is curved.

At each point A, if we want to compute the distance between A and a neighboring point B, we cannot apply Pythagoras theorem unless the local coordinate axes are orthonormal. In general we have to apply Al-Kashi's theorem,[1] which generalizes Pythagoras

---

[1] Al-Kashi (c. 1380–1429) was a Persian mathematician, born in Kashan, Iran. He discovered how, in Euclidean geometry, to adapt to any triangle Pythagoras theorem, which states that in a right triangle, whose shorter sides are a and b and the hypotenuse is c, we have

$$c^2 = a^2 + b^2$$

Al-Kashi's theorem says that, *for any triangle*, if the angle between the two shorter sides is θ, then we have

$$c^2 = a^2 + b^2 - 2ab\cos\theta$$

taking into account the cosine of the angle between the coordinate axes. We may also have to correct for units that are not unit distances on the axes. Yet the page is intrinsically flat, be it rolled or not in the embedding 3D Euclidean space. It is easy to find a set of coordinates Y's that will transform equation (1) into Pythagoras theorem. On the pages of primary school notebooks, they are usually shown. It doesn't disturb us to look at them, interpret them, and use them to locate a point, even when the page is furled.

Our ultimate mathematical goal, concerning the geometry of space-time in general relativity,[2] matches closely the question we addressed in the previous lecture of whether there is a real gravitational field or the apparent gravitational field is just due to an artifact of funny space-time coordinates. For instance, in figure 4 of lecture 1, the curvilinear coordinates were due to the accelerated frame we were using, not to tidal forces. The space-time was intrinsically flat. So we want to tackle the mathematical question:

Given the metric of a space-time as in equation (1), is the space-time really flat or not? Or, to put it another way, are there tidal forces or not? [3]

The mathematical question is a hard one. It will keep us busy during this lecture and the next.

As said in the introduction, we will first consider the question in a Riemannian geometry, where distance is defined locally and is always positive.

However, before we come to that, we need to get better acquainted with tensors. We began to talk about them in lecture 1. We introduced the basic contravariant and covariant transformation rules. In this lecture, we want to give a more formal presentation of tensors.

---

[2] An even more ultimate goal is to link the geometry of space-time with the distribution of masses and related concepts in space. These will be the Einstein field equations, which are the topic of lecture 9.

[3] If there are tidal forces, it is truly not flat in the sense of the geometry of space-time in general relativity.

Scalars and vectors are special cases of tensors. Now we are interested in the general category of tensors.

Scalar, Vector, and Tensor Fields

For us, tensors are indexed collections of values that depend on coordinate systems. Moreover they transform according to certain rules when we go from one coordinate system to another.

We are going to be interested in spaces such that at every point P of the space – the point P being located by its coordinates X in some system – there may be some physical quantities associated with that point. Such a function, which to each point of a space associates a thing (a scalar, a vector, a tensor, etc.), is called a *field* (respectively a scalar field, a vector field, and tensor field, etc.). The things or quantities that will interest us will be tensors. There will also be all kinds of quantities that will not be tensors. However we will mostly be interested in tensor fields.

The simplest kind of tensor field is a *scalar field* $S(X)$. It is a function that to every point in space associates a number, and everybody, no matter what coordinate system they use, agrees on the value of that scalar. So the transformation property in going, let's say, from the X^m coordinates to the Y^m coordinates is simply that the value of S at a given point P doesn't change.

We could use extremely cumbersome notations to express this fact in the most unambiguous way. But we will simply denote it

$$S'(Y) = S(X) \tag{2}$$

The right-hand side and the left-hand side denote *the value of the same field at the same point* P, one in the Y-system, the other in the X-system. The multidimensional quantity Y is the coordinates of P in the Y-system, while X is the coordinates of P in the X-system. For convenience, we add a prime to S when we talk of its value at P using the Y-coordinates. We saw in lecture 1 that sometimes we do without the prime sign (for instance, when we invoked the chain rule), but here we will keep it for the sake of clarity. With practice, equation (2) will become clear and unambiguous.

Remember that not all functions mapping the space onto the real numbers are scalar fields. They must also not change under a change of coordinates. For instance, when looking at a vector field in some coordinate system, if we decided to look only at its first component, it would not be a scalar field.

Let's represent, on a two-dimensional surface, the X coordinate system. Now, to avoid confusion, let's not embed the surface in any larger Euclidean space. But it can be truly curved.

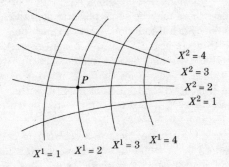

Figure 2: Curvilinear coordinates X's on a possibly curved surface.

Any point P of the surface can be located knowing the values of its two coordinates X^1 and X^2 in the X-system. Pay attention to the fact that we placed the indices of the coordinates upstairs, in other words, for the time being, we use superscripts.

Of course, we could think of a higher-dimensional space. There would then be more coordinates. Globally, we denote them X^m.

Now on the same space, there could be another coordinate system, a Y-system, to locate points, as shown in figure 3. In our figure, the point P has coordinates $(2, 2)$ in the X-system and $(5, 3)$ in the Y-system. Of course, these coordinates don't have to be integers. They can take their values in the set of real numbers.

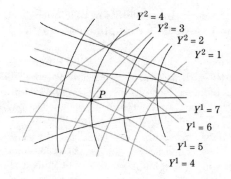

Figure 3: Second coordinates, the Y-system, on the surface.

What is important to note is that at any point P, there are two collections of coordinates: X^m and Y^m. The X^m and Y^m are related. At any point P, each coordinate X^m is a function of all the Y^m. And conversely. We write it this way:

$$X^m = X^m(Y) \qquad (3a)$$

$$Y^m = Y^m(X) \qquad (3b)$$

Equation (3a) is a coordinate transformation, and equation (3b) is its inverse. They can be pretty complicated, as long as they are one-to-one. We will also assume that the functions defined by equations (3a) and (3b) are continuous, and that we can differentiate them when need be, but nothing more.

Scalar fields transform trivially. If you know the value of S at a point P, you know it no matter what coordinate system you use.

Next are vectors. For us they come in two flavors. There are contravariant vectors, which we denote with an upstairs index V^m. And there are covariant vectors, which we denote with a downstairs index V_m. We spoke about them in the last lecture. Now we are going to delve deeper into their geometrical interpretation. What does it mean intuitively to be contravariant or to be covariant?

Geometric Interpretation of Contravariant and Covariant Components of a Vector

In this section and the mathematical interlude that follows, in order to distinguish as clearly as possible vectors from numbers, we shall use boldface for vectors and normal-face for numbers. In the subsequent sections on tensor mathematics and the following, however, we will revert to normal-face for everything. Even though boldface for vectors has advantages, it comes at the price of more cluttered equations. And equations of general relativity are already complicated enough!

Let's consider a coordinate system, and draw its axes as straight lines because we are not interested at the moment in the fact that the coordinates may be curved and may vary in direction from place to place. We could also think of them *locally*, where every manifold is approximately flat (a smooth surface, locally, is like a plane) and every coordinate system is formed of approximately straight lines, or surfaces if we are in more than two dimensions.

We are mostly concerned with the fact that the coordinate axes may not be perpendicular, and with the implications of that non-perpendicularity of these coordinates. Furthermore the distance between two axes, say $X^1 = 0$ and $X^1 = 1$, is not necessarily 1. The values of the coordinates are just *numerical labels*, which don't directly refer to distances.

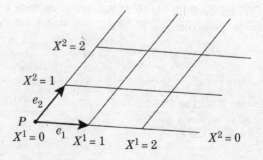

Figure 4: Coordinate system X. For simplicity we placed the point P under consideration at the origin of the system.

Now let's introduce some ordinary vectors pointing along the co-ordinates axes. On our two-dimensional surface, we introduce two vectors, e_1 and e_2, as shown in figure 4.

If we had three dimensions, there would be a third vector e_3 sticking out of the page, possibly slanted. We can label these vectors e_i. As the index i goes from 1 to the number of dimensions, the geometric vectors e_i's correspond to the various directions of the coordinate system.

Next in our geometric explanation of contravariant and covariant components of vectors, we consider an arbitrary ordinary vector V; see figure 5.

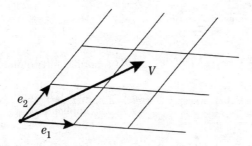

Figure 5: Vector V.

We have known since high school that the vector V can be expanded into a linear combination of the e_i's:

$$V = V^1 e_1 + V^2 e_2 + V^3 e_3 \qquad (4)$$

On the right-hand side of this formula, the quantities that are vectors are the e_i's. The V^i's are just a collection of numbers. As we explained in lecture 1, they are the *contravariant* components of the vector V in the e_i basis.

In summary, the contravariant components are the expansion co-efficients of V, i.e., the numbers that we have to put in front of the three vectors e_1, e_2, and e_3 to express a given vector as a sum of vectors colinear to those of the basis. This jibes with what we have said previously: the most usual vectors (used for position, translation, velocity, etc.) are contravariant vectors.

Earlier, I said that in Riemannian geometry a vector is just a vector – neither contravariant nor covariant – but that it has contravariant and covariant components. Our next job is to understand what the covariant components of the same vector V mean.

To take a peep into what's coming: to the vector V, we will attach *another* collection of numbers, this time denoted V_j's. They won't be the numbers to put in front of the units vectors e_1 and e_2 to construct a linear combination equal to V. They will be other things.

But first, a short interlude.

Mathematical Interlude: Dot Product of Two Vectors

Let's recall the elementary concept of the dot product between two vectors. I will keep it simple, assuming that you have seen it before (for instance, on page 27 of volume 1 of TTM).

Given any two vectors V and W, their dot product is defined as follows:

$$V \cdot W = |V|\,|W|\,\cos\theta \tag{5}$$

where $|V|$ and $|W|$ stand for the lengths of the vectors V and W, and θ is the angle between the two vectors. For example, if $|V|$ and $|W|$ are pointing in the same direction ($\theta = 0$), then the dot product is just the product of their lengths. If on the other hand they point in opposite directions ($\theta = \pi$), then the dot product is minus the product of their lengths. If $|V|$ and $|W|$ are orthogonal ($\theta = \pi/2$), their dot product is zero.

If we have an orthonormal basis at our disposal, then if V has the components V^1, V^2, ..., V^N in that basis and W has the components W^1, W^2, ..., W^N, where N is the dimension of the space, we know that the dot product also has the simple expression

$$V \cdot W = V^1 W^1 + V^2 W^2 + \ldots + V^N W^N \tag{6}$$

Exercise 1: Prove that, in an orthonormal basis, equation (5) is equivalent to equation (6).

Hint: Do it in two dimensions. Then – it is slightly more involved – we encourage you to try to do it in any dimension.

End of interlude.

We shall consider the dot products of V with the e_i's. Note that $(V \cdot e_1)$ is a number, whereas e_1 is a vector. For that reason the dot product is also called the *scalar product*.

By definition, the numbers $(V \cdot e_1)$, $(V \cdot e_2)$, ..., $(V \cdot e_N)$ are called the *covariant components* of the vector V. We denote them with subscripts:

$$V_i = V \cdot e_i \tag{7}$$

Andy: *Hey Lenny, when I went to high school and learned about vectors, the teacher never said covariant or contravariant components. Just plain old components. What gives?*

Lenny: *Yeah, that's because the teacher was using ordinary Cartesian coordinates. I'll explain:*

Cartesian coordinates are perpendicular to each other, and the e_i's are unit vectors. In that case the contravariant and covariant components are exactly the same. But if the coordinates are more general – for example, if they intersect at peculiar angles – then they won't be the same at all.

Andy: *This is crazy, Lenny. Why don't we just use the good old high school Cartesian coordinates and avoid all this complexity?*

Lenny: *Good question. You tell me.*

Andy: *Oh yeah, I forgot. In curved spaces there are no Cartesian coordinates.*

Lenny: *Yup.*

Let's see how we can relate the contravariant components V^m and the covariant components V_n. To reach that goal, we take the dot product of each side of equation (4) with e_n. This yields

$$V \cdot e_n = V^m e_m \cdot e_n \qquad (8)$$

The left-hand side, $V \cdot e_n$, is what we just defined as V_n. But the quantities $e_m \cdot e_n$ are something new. Note that they have two lower indices, which might lead us to expect that they are the components of a tensor of some sort. In fact, we will see that $e_m \cdot e_n$ turns out to be the metric tensor, expressed in the e_i's basis.

Pay attention also to the fact that on the right-hand side of equation (8) there is a sum to be done. The index m is a dummy index, to be summed over. This simplification of our work on vectors and their various components is a nice aftereffect of the summation convention.

Let's see how this connection between $e_m \cdot e_n$ and the metric tensor comes about.

The squared length of a vector is the dot product of the vector with itself. Let's calculate the length of V. Using twice equation (4) we write (for the squared length)

$$V \cdot V = V^m e_m \cdot V^n e_n \qquad (9)$$

We must use two different indices m and n. Indeed, recall that, in the implicit summation formula $V^m e_m$, the symbol m is only a dummy index. So in order not to mix things up, we use another dummy index n for the second expression of V. If you are not yet totally at ease with Einstein summation convention, remember that, written explicitly, the right-hand side of equation (9) means nothing more than

$$(V^1 e_1 + V^2 e_2 + V^3 e_3) \cdot (V^1 e_1 + V^2 e_2 + V^3 e_3)$$

But now, using the distributive property of the dot product, the right-hand side of equation (9) can be reorganized as

$$V \cdot V = V^m V^n (e_m \cdot e_n) \qquad (10)$$

The quantity $e_m \cdot e_n$ we call g_{mn}. So equation (10) becomes

$$V \cdot V = V^m V^n g_{mn} \tag{11}$$

This is exactly what the metric tensor is supposed to do; namely it tells us how to compute the square of the length – and therefore the length – of a vector. The vector could be, for instance, a small displacement dX.

In the following, in order not to clutter notations, we won't use boldface to denote vectors when we talk about dX's. And eventually, beginning with the next section, "Tensor Mathematics," we will altogether get rid of boldface for vectors.

In the case of a vector dX, equation (11) would be the computation of the square of the length of a little interval between two neighboring points

$$dX \cdot dX = dX^m dX^n g_{mn}$$

which is written more customarily

$$dX \cdot dX = g_{mn}\, dX^m dX^n \tag{12}$$

We now have a better understanding of the difference between covariant and contravariant indices, that is to say the covariant and contravariant components of a vector:

Contravariant components are the coefficients we use to construct a vector V out of the basis vectors. Covariant components are the dot products of V with the basis vectors.

The two types of components describe different geometric things. They would, however, be the same if we were speaking of ordinary Cartesian coordinates – meaning by that a basis made of vectors mutually orthogonal and each of unit length.

We inserted that discussion in order to give the reader some geometric idea of what covariant and contravariant mean and also what the metric tensor is. For a given collection of basis vectors e_i's and a given vector V, we summarize all this in the next box.

We shall also include another important formula.

There is indeed a very important equation relating the contravariant components of a vector with its covariant components. It uses the metric tensor g_{mn}. The relation is

$$V_n = g_{nm} V^m \tag{13}$$

We leave it to the reader to prove this. We will also meet the relation in the other direction with the help of the twice contravariant form of the metric, namely the tensor g^{nm}. But this, we will see later.

For the time being, let's recapitulate the important relations we established.

$$
\begin{aligned}
\boldsymbol{V} &= V^m \boldsymbol{e}_m \\
V_n &= \boldsymbol{V} \cdot \boldsymbol{e}_n \\
g_{mn} &= \boldsymbol{e}_m \cdot \boldsymbol{e}_n \\
V_n &= g_{nm} V^m
\end{aligned}
\tag{14}
$$

These relations are essential. We will make frequent use of them in the construction of the theory of general relativity.

Let's just make one more comment about the case when the coordinates axes are Cartesian coordinates from an orthonormal basis. Then, as we saw, the contravariant and the covariant components of $\boldsymbol{V}$ are the same, and the metric tensor is the unit matrix. Let's stress that this means that the basis vectors are perpendicular and of unit length.

Indeed, they could be orthogonal without being of unit length. In polar coordinates (see figure 14 of lecture 1), the basis vectors at any point P on the sphere are orthogonal, but they are not all of unit length. The longitudinal basis vector has a length that depends on the latitude. We have to use a coefficient equal to the cosine of the latitude. That is why, on the sphere of radius one, to compute the square of the length of an element dS, we can use Pythagoras theorem, but we must add $d\theta^2$ and $\cos^2\theta \; d\phi^2$ (see formula (16) of lecture 1).

We now come to the tensor mathematics that we will need throughout the rest of the book.

Tensor Mathematics

As we have said now several times, and we'll say it again, tensors are objects that are characterized by the way they transform under coordinate transformations. Let's just review quickly what we have found, and then go further. The transformation properties of the contravariant and covariant components of a vector were given in equations (24a) and (24b) of lecture 1. I repeat them here, with new labels.

Contravariant components

$$(V')^m = \frac{\partial Y^m}{\partial X^p} \; V^p \qquad (15a)$$

Covariant components

$$(W')_m = \frac{\partial X^p}{\partial Y^m} \; W_p \qquad (15b)$$

Let's go to tensors of higher rank. A tensor of higher rank simply means a tensor with more indices. Again, for the sake of pedagogy and completeness in this second lecture, there is some overlap with what we said at the end of the first lecture.

We start with a tensor of rank 2, with one contravariant index and one covariant index. It is a mathematical "thing" represented in a given basis by a collection of numbers.[4] These numbers are indexed with two indices. Furthermore in another basis the same "thing" is represented by another collection of numbers and the two collections satisfy specific transformation rules related to the relationship between the two bases. Let's consider the tensor in a Y basis, that is to say, a Y-coordinate system. We denote it

$$(W')^m{}_n$$

The simplest example of such a thing would be the outer product of two vectors, one with a contravariant index and one with a covariant index. By "outer product of the vectors" we mean the collection of all the products of components.[5] What makes the thing a tensor is its transformation property. So let's write it

$$(W')^m{}_n = \frac{\partial Y^m}{\partial X^p} \frac{\partial X^q}{\partial Y^n} W^p{}_q \qquad (16)$$

This tells us how a tensor of rank 2, with one contravariant and one covariant index, transforms. For each index on the left-hand side, there must be a $\partial Y/\partial X$ or a $\partial X/\partial Y$ on the right-hand side. You simply track where the indices go.

Let's do another example of a tensor of rank 2 with two covariant indices:

$$(W')_{mn}$$

How does it transform? By now you should begin to be able to write it mechanically

$$(W')_{mn} = \frac{\partial X^p}{\partial Y^m} \frac{\partial X^q}{\partial Y^n} W_{pq} \qquad (17)$$

These rules are very general. If you take a tensor with any number of indices, the pattern is always the same. To express the transformation rules from an unprimed system X to a primed system Y, you introduce partial derivatives, in one sense or the other as we did, on the right-hand side, and you sum over repeated indices.

---

[4]A tensor does have a geometric representation that by definition doesn't depend on the basis used. But in this course, we won't spend much time on this aspect of tensors. For us, in general relativity, it is the transformation properties of their components that are essential.

[5]The outer product is sometimes called the tensor product.

Notice one important property of tensors. If they are zero in one
frame, they are necessarily zero in any frame. This is obvious
for scalars: if a scalar is 0 in one frame, it is 0 in every frame,
because its value depends only on the geometric point where it
is measured, not the coordinates of that point. Now suppose a
vector V is zero in some frame, let's say the X-frame. To say that
V is zero doesn't mean that some component is equal to zero, it
means all of its components are zero. Equation (15a) or equation
(15b) then show that they are all going to be zero in any frame.

Likewise with any tensor, if all of its components are null in one
frame, that is, in one coordinate system, then all of its components are null in every frame.

That has an important and very useful consequence: once we have
written down an equation equating two tensors in one frame, for
instance

$$T^{lmn}_{pqr} = U^{lmn}_{pqr}$$

it can be rewritten

$$T^{lmn}_{pqr} - U^{lmn}_{pqr} = 0$$

Thus, considering that $T - U$ is still a tensor (see the next section,
"Tensor Algebra"), we see that

if two tensors are equal in one frame, they are equal in any frame.

That is the basic value of tensors. They allow you to express equations of various kinds, equations of motion, equations of whatever
you are working on, in a form where the same exact equation will
be true in any coordinate system. That is of course a deep advan-
tage to thinking about tensors.

We will also meet and use extensively other objects that will not
be tensors. Unfortunately it will be possible for them to be zero
in some frames and not zero in other frames. This will make life
a bit more complicated for us, but we will see how to deal with it.

Tensors have a certain invariance to them. Their components
are not invariant. They change from one frame to another. But
the statement that a tensor is equal to another tensor is frame-
independent. Incidentally, when you write a tensor equation, the

components have to match. It doesn't make sense to write an equation like W_q^p, where p is contravariant and q covariant, equals T^{pq}, where both indices are contravariant. Of course, you can write whatever you like, but if, let's say in one coordinate system, the equation $W_q^p = T^{pq}$ happened to be true (for all pairs p, q these are only numbers after all, so it is not meaningless), then it would usually not be true in another. So normally we wouldn't write equations like that.

One more point concerning vectors and higher-rank tensors: in Euclidean geometry, or in non-Euclidean geometry with a positive definite distance, for $V = W$ to be true, it is necessary and sufficient that the magnitude of $V - W$ be equal to zero.

But this statement is not true in the Minkowski geometry of relativity, where the proper distance between two events may be zero without them being the same event. The magnitude of a vector and the vector itself are two different things. The magnitude of a vector is a scalar, whereas the vector is a complex object. It has components. It points in a direction. To say that two vectors are equal means that their magnitudes are the same and their directions are the same. A tensor of higher rank is yet a more complicated object, which points in several directions. It has some aspect of it that points in one direction and some aspects that point in other directions. We will talk a bit about their geometry. But for the moment we define them by their transformation properties.

The next topic in tensor mathematics is operations on tensors. It usually bears the specific name of *tensor algebra*.

Tensor Algebra

What can we do with tensors that will produce new tensors? We are not interested at this point in operations we can do with tensors that produce other kinds of objects that are not tensors. We are interested in the operations we can do with tensors that will produce new tensors. That way we will be able to build equations having the very useful feature that they are frame-independent.

First of all we can multiply a tensor by a number. The result will still be a tensor. That rule is obvious and we don't need to spend time on it.

We shall examine three additional algebraic operations.

1. **Addition of tensors**. We can add two tensors of the same type, that is, of the same rank and the same numbers of contravariant and covariant indices. Addition of course also includes subtraction. If you multiply a tensor by a negative number and then add it, you are doing a subtraction.

2. **Multiplication of tensors**. We can multiply any pair of tensors to make another tensor.

3. **Contraction of a tensor**. From certain tensors we can produce tensors of lower rank.

Adding tensors. You only add tensors if their indices match and are of the same kind. For example, if you have a tensor

$$T = T^{m\cdots}_{\ldots p}$$

with a collection of upstairs contravariant indices and a collection of downstairs covariant indices, and you have another tensor *of the same kind*

$$S = S^{m\cdots}_{\ldots p}$$

in other words their indices match exactly, then you are allowed to add them and construct a new tensor, which we can denote

$$T + S$$

It is constructed in the obvious way: each component of the sum

$$(T + S)^{m\cdots}_{\ldots p}$$

is just the sum of the corresponding components of T and S. It is obvious too to check that $T + S$ transforms as a tensor with the same rules as T and S. The same is true of $T - S$. It is a tensor. This is the basis for saying that tensor equations are the same in every reference frame, because $T - S = 0$ is a tensor equation.

Multiplication of tensors. Unlike addition, multiplication of
tensors can be done with tensors of different rank and type. The
rank of a tensor is its number of indices. We know that the two
types, for each index, are contravariant or covariant. We can multiply T^l_{mn} by S^p_q. The tensor multiplication being not much more
than the multiplication of components and of the number of indices, we will get a tensor of the form P^{lp}_{mnq}.

Let's see again the simple example we already met: the tensor
multiplication, also called *tensor product*, of two vectors. Suppose
V^m is a vector with a contravariant index. Let's multiply it by
a vector W_n with a covariant index. This produces a tensor with
one upstairs index m and one downstairs index n:

$$V^m W_n = T^m_n \qquad (18)$$

A tensor is a set of values indexed by zero (in the case of a scalar),
one (in the case of a vector), or several indices. This tensor T of
equation (18) is a set of values (which, as we said many times,
depend on the coordinate system in which we look at it) indexed
by two indices m and n, respectively of contravariant and covariant type. It is a tensor of rank 2, contravariant in one index and
covariant in the other.

We could have done the multiplication with some other vector
X^n. This would have produced some other tensor:

$$V^m X^n = U^{mn} \qquad (19)$$

The tensor product is sometimes denoted with the sign $\otimes$. Equations (18) and (19) would then be written

$$V^m \otimes W_n = T^m_n$$
$$V^m \otimes X^n = U^{mn}$$

In this book we denote the tensor product by just writing the
tensors next to each other.

The tensor product of two vectors generalizes to the product of
any tensors. We produce a tensor of higher rank by just juxta-
posing somehow all the components of the multiplicands.

How many components does $V^m X^n$ have? Since we are going to
work mostly with 4-vectors in space-time, let's take V and X to
be both 4-vectors. Each is a tensor of rank 1 with a contravariant
index. Their tensor product U is a tensor of rank 2. It has 16 independent components, *each of which is the ordinary multiplication
of two numbers*:

$$U^{11} = V^1 X^1, \ U^{12} = V^1 X^2, \ U^{13} = V^1 X^3, \ \dots$$
$$\dots \ U^{43} = V^4 X^3, \ U^{44} = V^4 X^4$$

Observe that the tensor product of two vectors *is not* their dot
product. We will see how the dot product of two vectors is related to tensor algebra in a moment. The dot product has only
one component, not 16, and you might suspect that it is a scalar.
You'd be right. It is a frame-independent number.

Typically the tensor product of two tensors is a tensor of different
rank than either one of the multiplicands. The only way you can
make a tensor of the same rank is for one of the factors to be a
scalar. A scalar is a tensor of rank 0. You can always multiply
a tensor by a scalar. Take any scalar S and multiply it by, say,
V^m. You get another tensor of rank 1, i.e., another vector. It is
simply V elongated by the value of S. But generally you get back
a tensor of higher rank with more indices obviously.

Contraction. Contraction is also an easy algebraic process. But
in order to prove that the contraction of a tensor leads to a tensor,
we need a small theorem. No mathematician would call it a theorem. They would at most call it a lemma. Here is what the lemma
says. Consider the following quantity:[6]

$$\frac{\partial X^b}{\partial Y^m} \frac{\partial Y^m}{\partial X^a} \tag{20}$$

Remember that the presence of m upstairs and downstairs means
implicitly that there is a sum to be performed over m. Expression (20) is the same as

$$\sum_m \frac{\partial X^b}{\partial Y^m} \frac{\partial Y^m}{\partial X^a} \tag{21}$$

---

[6]We begin to use also letters a, b, c, etc. for indices because there just
aren't enough letters in the m range or the p range for our needs.

What is the object in expression (20) or (21)? Do you recognize what it is? It is the change in X^b when we change Y^m a little bit, times the change in Y^m when you change X^a a little bit, summed over m. That is, we change Y^1 a little bit, then we change Y^2 a little bit, etc. What is expression (21) supposed to be?

Let's go over it in detail. Instead of X^b, consider any function F. Suppose F depends on $(Y^1, Y^2, \ldots, Y^M)$ and each Y^m depends on X^a. Then, from elementary calculus, the quantity

$$\frac{\partial F}{\partial Y^m} \frac{\partial Y^m}{\partial X^a}$$

is nothing more than the partial derivative of F with respect to X^a (partial because there can be other X^n's on which the Y^m's depend). That is

$$\frac{\partial F}{\partial Y^m} \frac{\partial Y^m}{\partial X^a} = \frac{\partial F}{\partial X^a}$$

What if F happens to be X^b? Well, there is nothing special in the formulas. We get

$$\frac{\partial X^b}{\partial Y^m} \frac{\partial Y^m}{\partial X^a} = \frac{\partial X^b}{\partial X^a}$$

What is $\partial X^b/\partial X^a$? It looks trivial. The X^n's are independent variables, so the partial derivative of one with respect to another is either 1, if they are the same, or 0 otherwise. So $\partial X^b/\partial X^a$ is the Kronecker-delta symbol. We shall denote it

$$\delta_a^b$$

Notice that we use an upper index and a lower index. We shall find out that δ_a^b itself also happens to be a tensor. That is a little weird because it is just a set of numbers. But it is a tensor with one contravariant and one covariant index.

Now that we have spelled out the little lemma we need in order to understand index contraction, let's do an example. Then we'll define contraction more generally.

Consider a tensor built out of two vectors, one with a contravariant index and the other with a covariant index:

$$T^m_n = V^m \ W_n \tag{22}$$

What contraction means is: *take any upper index and any lower index and set them to be the same and sum over them.* In other words, take

$$V^m \ W_m \tag{23}$$

Expression (23) means $V^1 W_1 + V^2 W_2 + V^3 W_3 + \ldots + V^M W_M$, if M is the dimension of the space we are working with. We have identified an upper index with a lower index. We are not allowed to do this with two upper indices, nor with two lower indices. But we can take an upper index and a lower index. Let's see how expression (23) transforms. For that, look at the transformation rule applied first to expression (22). We already know that it is a tensor. Here is how it transforms:[7]

$$(V^m \ W_n)' = \frac{\partial Y^m}{\partial X^a} \ \frac{\partial X^b}{\partial Y^n} \ (V^a \ W_b) \tag{24}$$

Equation (24) is the transformation property of the tensor T^m_n, which has one index upstairs and one index downstairs.

Now let $m = n$ and contract the indices by identifying the upper and the lower index *and sum over them.* On the left-hand side we get

$$(V^m \ W_m)'$$

How many indices does it have? Zero. So the contraction of $V^m \ W_n$ did create another tensor, namely a scalar.

We can check what equation (24) says. It should confirm that $(V^m \ W_m)'$ is the same as $V^m \ W_m$. Now our little lemma comes in handy. On the right-hand side of (24), when we set $m = n$ and

---

[7] We write $(V^m \ W_n)'$, but we could also write $(V^m)' \ (W_n)'$, because we know that they are the same. Indeed, *that is what we mean when we say that the outer product of two vectors forms a tensor*: we mean that we can take the collection of products of their components in any coordinate system. Calculated in any two systems, $(V^m)' \ (W_n)'$ and $V^m \ W_n$ will be related by equation (24).

sum over m, the sum of the products of partial derivatives is δ_a^b. So the right-hand side is $V^a\,W_a$. But a or m are only dummy indices, therefore equation (24) says indeed that

$$(V^m\,W_m)' = V^m\,W_m$$

It is easy to prove, and the reader is encouraged to do it, that if you take any tensor with a bunch of indices, any number of indices upstairs and downstairs,

$$T^{nmr}_{pqs} \tag{25}$$

and you contract a pair of them (one contravariant and one co-variant), say r and q, you get

$$T^{nmr}_{prs} \tag{26}$$

where the expression implicitly means a sum of components over r, and this is a new tensor.

Notice that the tensor of expression (25) has six indices, whereas the tensor of expression (26) has only four.

Notice also two more things:

1. If we looked at $V^m\,W^n$, we would be dealing with a tensor that cannot be contracted. The analog of equation (24) would involve

$$\frac{\partial Y^m}{\partial X^a}\,\frac{\partial Y^n}{\partial X^b}$$

   This quantity doesn't become the Kronecker-delta when we set $m = n$ and sum over it. The sum $\sum_m (V^m)'\,(W^m)'$ would not be equal to $\sum_m V^m\,W^m$.

2. The dot product of two vectors V and W is the contraction of the tensor $V^m\,W_n$. But in that case one vector must have a contravariant index, and the other a covariant index.

In other words, contraction is the generalization of the dot product, also called inner product, of two vectors. We are going to deal with inner products as soon as we work again with the metric tensor.

More on the Metric Tensor

Of all the tensors in Riemannian geometry, the metric tensor is the most important. In equations (14) we described its construction in terms of the basis vectors e_m's:

$$g_{mn} = e_m.e_n$$

Let's now define it on its own terms abstractly. These are things we have already covered before, but let's do them again now that we have more practice with tensors.

To define the metric tensor, consider a differential element dX^m that represents the components of a displacement vector dX located at a point P as in figure 6. And we consider an infinitesimal displacement, which we call dX.

Figure 6: Displacement vector dX.

The contravariant components of dX are the coefficients of the vector dX in the expansion given by equation (4). In the case of three dimensions,

$$dX = dX^1 e_1 + dX^2 e_2 + dX^3 e_3 \qquad (27)$$

What is the length of that displacement vector? To answer that, we need to know more about the geometry – in particular we need to know the metric tensor $g_{mn}(X)$ and how it varies from place to place. Writing the length of dX as dS, the generalization of the Pythagoras theorem is

$$dS^2 = g_{mn}(X) \, dX^m dX^n \qquad (28)$$

Mathematical Interlude: The Metric is a Symmetric Tensor

Any rank-2 tensor – let's call it T – can be written as a sum of a symmetric tensor and an antisymmetric tensor,

$$T_{mn} = S_{mn} + A_{mn}$$

where the symmetric part satisfies

$$S_{mn} = S_{nm}$$

and the antisymmetric part satisfies

$$A_{mn} = -A_{nm}$$

It follows that

$$dS^2 = S_{mn}\, dX^m dX^n + A_{mn}\, dX^m dX^n$$

Notice that because A is antisymmetric, the second term will always be zero. Thus without any loss of generality, we may assume that the metric tensor is symmetric:

$$g_{mn} = g_{nm}$$

The metric, like any other rank-2 tensor, can be displayed as a matrix with N^2 components. For example, if the space is four-dimensional, the metric would be a 4×4 matrix with 16 components, but because it is symmetric there are only 10 independent components, as shown in figure 7.

	X^1	X^2	X^3	X^4
X^1	✓	✓	✓	✓
X^2		✓	✓	✓
X^3			✓	✓
X^4				✓

Figure 7: Independent components in g_{mn}.

Similarly in a three-dimensional space there would be six independent components in g_{mn}. In two dimensions there would be three.

End of interlude

So far we haven't proved that g_{mn} is a tensor. I called it the metric tensor, but let's now prove that it is indeed such an object. The basic guiding principle is that the length of a vector is a scalar, and that everybody agrees on that length. People using different coordinate systems won't agree on the components of dX (see figure 6), but they will agree on its length. Let's write again the length of dX, or rather its square:

$$dS^2 = g_{mn}(X) \ dX^m \ dX^n \tag{29}$$

Now let's go from the X-coordinates to the Y-coordinates. Because dS^2 is invariant, the following holds:

$$g_{mn}(X) \ dX^m \ dX^n = g'_{pq}(Y) \ dY^p \ dY^q \tag{30}$$

Then let's use this elementary calculus fact:

$$dX^m = \frac{\partial X^m}{\partial Y^p} \ dY^p \tag{31}$$

Plug expression (31) for dX^m and for dX^n into (30). We get

$$g_{mn}(X) \ \frac{\partial X^m}{\partial Y^p} \ \frac{\partial X^n}{\partial Y^q} \ dY^p \ dY^q = g'_{pq}(Y) \ dY^p \ dY^q \tag{32}$$

The two sides of equation (32) are expressions of the same quadratic form in the dY^p's. That can only be true if the coefficients are the same. Therefore we have established the following transformation property:

$$g'_{pq}(Y) = g_{mn}(X) \ \frac{\partial X^m}{\partial Y^p} \ \frac{\partial X^n}{\partial Y^q} \tag{33}$$

This is precisely the transformation property of a tensor with two covariant indices. So we discovered that the metric tensor is indeed really a tensor. It transforms as a tensor. This will have numerous applications.

The metric tensor has two lower indices because it multiplies the differential displacements dX^m's in equation (29), which have upper indices.

The metric tensor can also be viewed as a matrix with m n indices. Remembering that $g_{ij} = g_{ji}$, it is the following matrix, which we still denote g_{mn},

$$g_{mn} = \begin{pmatrix} g_{11} & g_{12} & g_{13} & g_{14} \\ g_{12} & g_{22} & g_{23} & g_{24} \\ g_{13} & g_{23} & g_{33} & g_{34} \\ g_{14} & g_{24} & g_{34} & g_{44} \end{pmatrix}$$

It is a symmetric matrix.

There is one more fact about this matrix, that is, about the tensor g_{mn} thought of as a matrix. It has eigenvalues. These eigenvalues are positive and never zero.

The reason that the eigenvalues are never zero is because a zero eigenvalue would correspond to a eigenvector of length zero. But there is no vector of length zero (unless of course its components are all zero). In Riemannian geometry every direction has a positive length associated with it.

The Matrix Inverse of the Metric

What do we know about matrices that are symmetric and whose eigenvalues are all nonzero? Answer: they have inverses. The matrix of the metric tensor – denoted g_{mn} or g for simplicity – has an inverse g^{-1} whose components are themselves the components of a tensor, albeit with contravariant elements. The components of g^{-1} are written g^{mn}.

In matrix terms the product of the matrices g and g^{-1} is the identity matrix. This is represented by the formal equation

$$g^{-1} \, g = \text{the unit matrix}$$

In terms of components, it takes the form

$$g_{mn} \, g^{np} = \delta_m^p \tag{34}$$

where δ_m^p is the identity matrix.

Equation (34) is the definition of the matrix inverse, but it is also a tensor equation. We've already seen that the Kronecker-delta δ_m^p is a tensor with one lower and one upper index. That's enough to prove that g^{np} is a tensor with two upper indices.

In fact the three tensors g_{mn}, g^{mn}, and g_m^n are really just a single tensor written in three forms[8] – the first with two covariant indices, the second with two contravariant indices, and the third with a covariant and a contravariant index.

The fact that there is a metric tensor with downstairs indices and a metric tensor with upstairs indices will play an important role.

So far everything we have seen on tensors was easy. It is essentially learning and getting accustomed to the notation.

This lecture was about tensor algebra. The next lecture deals with tensor calculus: in particular with the dark art of parallel transporting tensors, differentiating tensors, and most importantly building a curvature tensor from the derivatives.

It is the curvature tensor that will tell us if a geometry is flat or curved, and in general relativity whether a gravitational field exerts tidal forces.

---

[8]Similarly, we saw that a rank-1 tensor, i.e., what we called in lecture 1 an abstract vector, has a contravariant form and a covariant form.

Lecture 3: Flatness and Curvature

Lenny: *Today, we shall study the difference between flat space and non-flat space.*

Andy: *I guess I know: we can always tile a plane with an infinite number of identical flat square tiles. And that's unrelated to the fact that we may, for some reason, use curvilinear coordinates. But we cannot tile in such a way the surface of the Earth.*

Lenny: *That's exactly right.*

Andy: *But the 3D space, we can always fill with identical cubes! I remember from my kindergarten days.*

Lenny: *Only locally. Globally, we're not sure what's the shape of the universe. We will talk about it in the next volume.*

Andy: *Oh, I get it. It's like: Think globally, act locally.*

Lenny: *Sort of.* This is the kind of gentle answer one makes to a suggestion that bears no relation to the subject. :-)

Introduction

General relativity has a reputation for being very difficult. I think the reason is that it *is* very difficult. It is calculation-intensive: symbols, indices, awe-inspiring equations. There are ways people have invented to express things in more condensed notations, but just learning them in itself is a task. Things like vierbeins, forms, spinors, and twistors and all sorts of other mathematical objects. You could call many of them just notational devices if you like. And they do simplify the equations. So I sometimes feel that in presenting these things the way I do, it is sort of like Maxwell[1] who wrote down every single equation of his set of Maxwell equations. At first he wrote twenty altogether. Now we write only four. We don't usually write all components of the equations. We put them together into vector notation and so forth. If we are smart, we can even avoid the indices by inventing symbols like del, curl, or Laplacian. The same thing could be done to some extent for general relativity. But in the end the computational techniques are unquestionably harder.

I will tend to downplay the computational side of things and concentrate on the principles. If you are really interested in doing the computations in general relativity, there are packages. You just put in the metric as a function of position, and the computer will spit out the various tensors that you ask it for: Riemann tensors, Ricci tensors, Einstein tensors, this kind of tensor, that kind of tensor. Then you can say, without even looking at the results: "Okay, please Mr. Computer, set the Einstein tensor equal to the energy momentum tensor and tell me what comes out." So, yes, computers can do a lot better than us.

General Relativity in Modern Physics

When I was a young physicist in the 1960s, general relativity was a bit of a backwater in theoretical physics. Partly this was because the technology for detecting the subtle non-Newtonian effects mostly did not yet exist. But also physicists like myself, who were interested in the fundamental aspects of the subject, had other fish to fry. Elementary particle physics – both theoretical and experimental – was in its golden age with new discoveries almost every year.

Things have changed since then. New technologies allowed new experiments and astronomical observations, which finally cleared any doubt about the correctness and importance of general relativity. It became clear that Einstein's theory was absolutely central

---

[1] James Clerk Maxwell (1831–1879), Scottish theoretical physicist.

to cosmology – the study of the origin and structure of the universe. Black holes were discovered at the centers of galaxies. It became urgent to have computational tools to numerically solve the equations of general relativity. That entailed a deeper understanding of those equations. The field of numerical relativity was born and flourished.

On a more theoretical side, string theory, which was originally designed to understand particle physics, provided new and powerful tools for analyzing gravity. Perhaps even more important were the clashes that were being discovered between quantum mechanics and general relativity. By the year 2000 there was no alternative: the dominant question for theoretical physicists was to understand how quantum physics and general relativity could be reconciled, or even unified.

There was more. One of the big surprises for me was the way the theoretical tools of gravity and quantum gravity found application to other fields, including condensed matter physics and quantum computer science. In short, general relativity is no longer a backwater. It is the mainstream.

Riemannian Geometry

This is the last lecture in which we will be studying Riemannian geometry as such, without really discussing gravity. In the next lecture we will really get into gravity. What do all these manipulations of tensors have to do with gravity? We already had a glimpse of the answer in the first lecture.

The problem of finding out whether there is a real gravitational field, as opposed to just some artifact of curvy coordinates, is mathematically identical to the problem of finding out if a certain geometry – characterized by its metric tensor – is flat or not.

Let's think of a two-dimensional space or "manifold" to begin with. That means a surface $\mathbb{S}$ where each point is located with two real coordinates X^1 and X^2, see figure 12 of lecture 1, where it is shown embedded in the usual 3D space for convenience.

We assume that $\mathbb{S}$ has a metric that defines infinitesimal lengths, the squares of which are given by

$$dS^2 = g_{mn}(X)\ dX^m\ dX^n \tag{1}$$

where m and n run over the indices $\{1,\ 2\}$.

A flat geometry is one for which all Euclid's axioms, including the famous fifth one called *Euclid's postulate*, are correct.[2] There are points, lines, parallel lines, distances, right angles, etc., all the stuff we learned in high school that is called Euclidean geometry. Moreover, the surface $\mathbb{S}$ – if we think of it embedded in 3D – is not necessarily a plane, but can be laid out on a plane without imposing any distortion, stretching, or compression, on its intrinsic geometry. Such surfaces in 3D are called *developable*. Cylinders and cones are examples.

A flat geometry is one where *we can find another system* of coordinates Y such that at any point P, now located by those Y-coordinates, the metric has a simple form

$$dS^2 = (dY^1)^2 + (dY^2)^2 \tag{2}$$

Such a transformation is always possible locally at any *given* point P, because locally any smooth surface is like a plane. But it is not always possible to find such a transformation globally over the whole surface. In other words, given an arbitrary metric tensor that varies from place to place – assuming it is smooth and has all the good differential properties – finding such coordinates doesn't always have a solution. And determining whether there exists one or not is in general a difficult problem.

The bad way to approach it is to search through all possible coordinate systems Y and see whether the transformed metric is the Kronecker matrix. This would take an infinite amount of time.

We need a better technique. The better technique is to search for a *diagnostic quantity*, built out of the metric and its derivatives, that we, or the computer, can calculate. If it is zero everywhere,

---

[2]It was called a *postulate* because for 2000 years people thought that it was "really true in reality," until their epistemology got more sophisticated.

then the space is flat. If it is nonzero at some location, this will tell us that the space shows some curvature there.

In the two-dimensional case, the diagnostic quantity that does the job is called the *Gaussian curvature*. More generally in higher dimensions, it is the *curvature tensor*. It's a bit of a slog, but it's worth the effort. Once we have mastered the curvature tensor, we have a very powerful tool, both for pure geometry and for understanding gravity.

What do we start with? We start with a space. And a space means, first of all, a number of dimensions. In Riemannian geometry the number of dimensions can be any positive integer. In principle you can even have a zero-dimensional space, but that is just a point! There isn't much to be said about the geometry of a point. So let's go to the next number of dimensions.

Andy: *Lenny, you don't seem to be well-versed in Leibniz's monads.*[3] *Actually a space consisting of only one point is fascinating! :-)*

Lenny: *Evidently Leibniz thought so – Newton not so much. :-(*

A one-dimensional space is either an infinite line or a closed curve – that is, a loop. If it is a closed curve, what is it intrinsically characterized by? One thing and only one: the total length of the curve. Every loop is equivalent to every other loop of the same length. In other words – just think about it for a moment – take a piece of rope that closes on itself to form a loop and has a certain length. Wiggle it or curve it in any kind of way, it can always still be mapped into or put on top of another piece of rope of exactly the same length. To a one-dimensional bug living on the loop, there is no more – just the length. All the bug can do is count the number of steps it takes to walk around the loop. For instance, the bug might make an initial mark someplace, then go around the loop till it comes back to the mark, and record the number

---

[3]Leibniz's monads didn't lead to any interesting understanding of the world. One then wonders why he devised such a weird concept. The reason is probably to be found in his invention of integral calculus, which met with fantastic success and rested on infinitesimal quantities. Monads are cousins of those. But, if infinitesimals – put on a firm footing in the nineteenth century – turned out to be very useful to this day, monads did not.

of steps that it took. That is the only thing the bug can say, or measure, about the loop.

In short, in one-dimensional spaces, there is no notion of curvature, only a notion of length. The reader might find this strange, because when we drive on a road – a one-dimensional space – there are straight sections and there are turns! That is correct, but one must understand that the notion of turns on a road is meaningful only if we consider the road embedded in a space of at least two dimensions, that is, a plane or surface, or a 3D space, etc.

Two-dimensional spaces are where things start to be more complicated – and more interesting. There are flat ones. And there are curved ones. A flat one is a plane. A curved one could be a sphere. It could be a space with bumps, the surface of the Earth including the mountains and the valleys. It could even have a weird topology, for instance the surface of a donut, also called a *torus*. You can poke another hole in the torus and make a torus with two holes, and so forth.

Things only get worse as the dimension increases. The sheer variety of different types of spaces in three-, four-, and higher-dimensional spaces is bewildering, but luckily we only need to know about a few simple cases.

Let us come to our main goal of finding a tensor that can distinguish whether a space is flat or curved by whether the tensor is zero or not. Why a tensor? Because flatness does not depend of the choice of coordinates. And tensors, if they are zero in one frame, then they are zero in all frames. The curvature tensor is traditionally denoted R in honor of Riemann. We will see that the curvature tensor is of rank 4, meaning it has four indices.

Gaussian Normal Coordinates

If an N-dimensional space is flat, we can choose coordinates in which the metric has the form

$$dS^2 = (dY^1)^2 + (dY^2)^2 + ... + (dY^N)^2 \qquad (3)$$

or, using the Kronecker-delta symbol,

$$dS^2 = \delta_{ij}\, dY^i dY^j \qquad (4)$$

To what extent can we force the metric to look like this by a choice of coordinates? In general the best we can do is to make the metric tensor be approximately δ_{ij} over a small region of space surrounding a point.

Here is a theorem that will be very useful to us:

At any given point P in the space, we can find a system of coordinates in which the metric is δ_{mn} to first order in small deviations from the point; see figure 1. In general, unless the space is flat, the attempt will fail beyond first order.

Such coordinates are called *Gaussian normal coordinates at the point P.* Here is how we proceed. We position ourselves at point P and we move along any first direction as straight as we possibly can. Later we will learn what is meant by "as straight as we possibly can"; it will mean along a geodesic. So you make as straight a curve as you can.

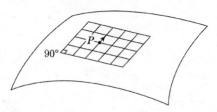

Figure 1: Displacement of length ΔS along the surface and along the tangent plane in the same direction. The coordinates are represented on the tangent plane at P. We could have represented them too – slightly curved – on the surface itself.

As an example, suppose that you are a little bug driving a tiny car on a two-dimensional surface. You move along the surface pointing your steering wheel straight ahead. That's what I mean by "as straight as you can."

That defines one coordinate axis. Then you come back to point P. You have some surveying tools to figure out which other directions make a right angle with the first line. On a two-dimensional surface there is only one other direction (in one sense or the other). In three dimensions, there is a whole plane. You go off in an orthogonal direction, again as straight as you can. That way you build a complete set of coordinates based on those directions.

The theorem says that at every point P of the surface, you can choose Gaussian normal coordinates such that, at that point whose coordinates are, say, X_0, we have

$$g_{mn}(X_0) = \delta_{mn} \tag{5}$$

You can do that in more than one way. If you found coordinates for which equation (5) is true, you can obviously rotate the coordinates. This will produce a different set of axes such that equation (5) in the new set is still true. In figure 1, think of pivoting the coordinate system around P.

The theorem says, furthermore, that at point P once you have chosen the directions, you can also choose the X's such that the *derivative* of any element of the metric tensor $g_{mn}(X)$ at that point with respect to any direction in space, X^r, can be set equal to zero:

$$\frac{\partial g_{mn}}{\partial X^r} = 0 \tag{6}$$

The proof is actually very simple. It is just a counting argument. You count how many independent variables you have, and how many constraints they must satisfy.

Equation (6) will be true, at a given point, only for the first derivatives. Unless the space is flat, the derivatives of higher order at that point won't be zero:

$$\frac{\partial^2 g_{mn}}{\partial X^r \partial X^s} \neq 0 \tag{7}$$

So, at a point, there is no content really in saying that the metric can be chosen to be, so to speak, flat-like. Up to the first derivatives included, that can always be done.

It is in the second derivatives of the metric tensor that the flatness or non-flatness of the space somehow starts to show up.

How do we prove it? As said, this is actually not hard. Let's do it. We set the point of interest, which we called P, of coordinates X_0, to be the origin:

$$X_0 = 0$$

Now suppose that we have some general metric and some coordinates Y in which the metric has some form that does not satisfy equations (6).

Let's look for some X's, which will be functions of the Y's, and choose them in the following way: at the place where $X = 0$, in other words at the origin, let's also assume that $Y = 0$. So the two sets of coordinates have the same origin. That means that X will start out just equal to Y plus something quadratic in Y:

$$X^m = Y^m + C_{nr}^m\, Y^n Y^r \tag{8}$$

plus some more complicated terms. We are simply expanding each X^m in powers of $Y^1, Y^2, \ldots, Y^N$, where N is the number of dimensions of the space.

How many such C_{nr}^m are there? Suppose we work in four dimensions. Then there are 10 distinct combinations $Y^n Y^r$, because $Y^n Y^r = Y^r Y^n$. For each n and r, we have four C_{nr}^m when m runs from 1 to 4. That means there are 40 independent coefficients. Now how many independent components of g are there? Answer: 10. So there are 40 equations (6). Finally we have reached 40 equations to solve for 40 unknowns. That allows us to be sure, at point P, not only that $g_{mn}(X) = \delta_{mn}$, but also that the derivatives of g_{mn} and δ_{mn} will match up to quadratic order.[4] It means that we will be able to solve the 40 equations (6), and, moreover, that we will fail to set the left-hand sides of equations (7) equal to zero.

To summarize: at any point P, a smooth space (or surface, or manifold) is locally flat. We can approximate it by its tangent

---

[4]It is easy to check that we are in a case where the 40 equations with 40 unknowns do lead to an existing and unique solution. The reader is invited to verify it in two dimensions.

space, as in figure 1. And we can construct coordinates X's such that P is located at the origin and the metric tensor has the form

$$g_{mn}(X) = \delta_{mn} + o(X) \tag{9}$$

where $o(X)$ represents terms of second order and higher. We interpret equation (9) as saying that the metric is locally Euclidean up to second order.

The fact that we cannot generally satisfy the equations to higher order demonstrates that generally spaces are not flat.

Our next goal is to learn to differentiate tensor fields with respect to position, so as to produce new tensors. This is a subtle business that will lead to the very important notion of the *covariant derivative* of a tensor.

Covariant Derivatives

To differentiate a tensor with respect to position, we could think: "Okay. Let's take the components of the tensor – for instance, contravariant components – and just differentiate them." That would yield a new collection of components – with one more index – which would be simply the derivatives of the components of the initial tensor. But we would run into a problem. Let's see what the problem is.

Think for instance of the derivatives of a vector. We could differentiate each component with respect to each direction. We can certainly do that. This would produce a two-dimensional collection of values. *But it would not be a tensor.* Here is why.

Consider a surface and a point P on it, figure 2. We have two sets of coordinates on the surface, coordinates X and coordinates Y.

If the space is flat, for X we just use ordinary flat Cartesian coordinates. Or if the space is curved, we use a set of Gaussian normal coordinates at P, that is, coordinates X that are locally, at P, as straight and orthogonal as possible, as we explained. And there is another set of coordinates Y, for instance the initial ones. For convenience, at P we chose X such that the X^2-axis is tangent to the Y^2-axis. Remember that we can rotate the Gaussian coordinates that we built so that they suit whatever purpose we have. So it's not a problem to make the X^2-axis parallel to the Y^2-axis.

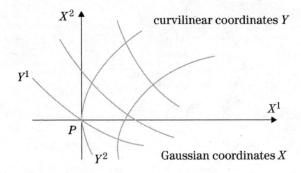

Figure 2: Surface viewed at P with Gaussian normal coordinates X, and arbitrary curvy coordinates Y.

Think of a vector field defined over the surface. The vector field is made of a different vector at every point. In order not to clutter the picture, these vectors are not shown on figure 2 yet. There is one at P and there are plenty around P – one at each point.

Before we ask how to differentiate a vector field, let's ask what it would mean for the vector field to be constant in space. We run into the following difficulty: because the space is curved, it becomes hard to compare the vector at one point with the vector at another point.

The coordinates X cannot be chosen to be everywhere flat. Then what exactly do we mean by saying that a vector at one point is equal to a vector at another point? It does not really mean anything because to compare a vector at P with a vector at Q, unless we have nice flat coordinates over the whole surface, there is no unique way to do that. Let's look at this in detail.

If the space is really flat, then we know what it means for a vector at one point P of the surface to be the same as a vector at another point Q of the surface; see figure 3. It means that they point in the same direction and have the same length. Therefore in the X coordinates they have the same components.

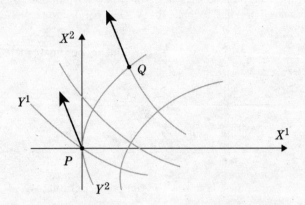

Figure 3: Two equal vectors, at P and at Q.

What about their components in the Y coordinate system?

To understand clearly what's going on, let's take a special case. Suppose that both vectors point vertically in the X-axes; see figure 4. In that case, V_P only has an X^2 component in the flat coordinates. And so does V_Q. They have the same components in the X-axes.

Are the components of V_P and V_Q in the Y-coordinates the same? Answer: No. Along the Y-axes, V_Q has a Y^1 component and a Y^2 component, while V_P only has a Y^2 component.

It's clear that even though the vectors V_P and V_Q are the same, their components are not the same in the Y coordinate system.

This will be true, be it for covariant components or for contravariant components: when we have curvilinear coordinates, we cannot easily judge the equality of two vectors at separated points.

Furthermore, to make matters worse – or more interesting – in curved space there are only curvilinear coordinates.

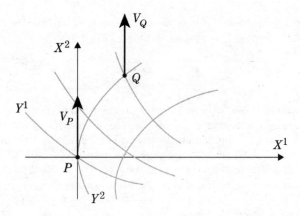

Figure 4: Two equal vertical vectors, at P and at Q. (We have represented a part of the grid in the Y-axes, in the vicinity of point P, but not in the X-axes. In the X-axes, the grid is approximately rectangular like in Euclidean geometry.)

Another way to say the same thing is that the derivative of the m-th component of V, with respect to the r-th direction of the coordinate system, might be zero in one coordinate system and not zero in the other coordinate system:

$$\frac{\partial V_m}{\partial X^r} = 0$$

$$\frac{\partial V'_m}{\partial Y^r} \neq 0$$

It might even be the case – as in figure 3 or 4 – that *all of the derivatives* of V_m are zero in one coordinate system and not zero in the other coordinate system. That will be because the coordinates are shifting, not because the vector is changing.

You see what we are driving at: the equation saying that the derivatives of a vector are all equal to zero may be true in one reference frame, but untrue in another. Therefore it cannot be a tensor equation. Let's highlight this fact:

The ordinary derivatives of the components of a vector with respect to the coordinates do not themselves form a tensor.

If they were components of a tensor, we would think of this quantity

$$T_{mr} = \frac{\partial V_m}{\partial X^r}$$

as a rank-2 tensor with an m index and an r index. But if it were a tensor, the following fact would have to be true: if T_{mr} is zero in one frame or one coordinate system, it is zero in every coordinate system. Yet it is simply not true with that T_{mr} – not because the vector may change from point to point, but, as we illustrated, because the orientation of the coordinates changes.

We need a better definition of the derivative of a vector than just differentiating its components. We need something that if it is zero in one frame is zero in every frame.

Here is how we will define the derivative of a vector. As a preliminary, notice that to define the derivative at a point P we only need to look at points in the vicinity of P. The first thing we do is to construct a set of Gaussian normal coordinates at point P. Remember: Gaussian normal coordinates are as straight as possible near P. They are well defined over the whole space, and they make up an approximately Euclidean system of coordinates *in the vicinity of* P. So we re-express all the vectors of the vector field in the new coordinates X that are locally flat Euclidean at P.

To follow the procedure geometrically, let's look again at two vectors of the vector field: the vector corresponding (or "attached") to point P and the other vector corresponding to a nearby point Q, as in figure 5. For clarity make the second vector also slightly different from the first. Then pretend that the Gaussian normal coordinates are really nice flat coordinates over the whole vicinity of P. Visually translate the vector V_Q so that its origin is the same as V_P's, and look at the difference between V_Q and V_P.

The components, in Gaussian normal coordinates, of the difference between V_Q and V_P are the kind of elements that we will use to define the derivatives of the vector field at P. For instance, if we were interested in the derivative along the PQ direction, it would approximately be the vector $V_Q - V_P$ divided by the small distance between P and Q. But we shall be interested in the derivatives of V along the X-axes.

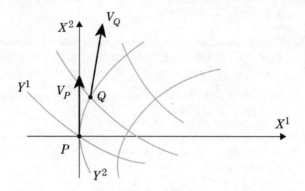

Figure 5: Two vectors corresponding to nearby points.

Now the derivatives $\partial V_m/\partial X^r$, in the Gaussian normal coordinates, *define* the derivatives of V at P.

Finally, if we want to work with our initial Y-coordinates, we take the double-indexed collection of partial derivatives produced by the differentiation in Gaussian normal coordinates at P. We treat it as a tensor; that is, we consider this collection as the components of a tensor in the X-coordinates. And we transform it back into the Y-coordinates using the tensor equations linking X and Y. This necessarily produces a tensor, since when transformed from any Y-coordinates into the X-coordinates, it gives the same thing.

When we look at the derivatives of a vector in the general coordinates Y, we will have the addition of two terms: one term because the vector may be changing and the other term because the coordinates may be changing. As we saw, the coordinates may shift, may be rotating out from under you, even if the vector doesn't change; see figures 3 and 4.

Before we look into these two terms, let's repeat the prescription as a methodical procedure.

We have a vector field V on a space equipped with any coordinate system Y. We want to calculate the derivative of V at a point P. Then we follow these steps:

1. Change coordinates to use Gaussian normal coordinates at P, let's call them X (notice that they are valid over the whole surface and approximately flat at P).

2. Differentiate V at P in the usual way, using the X-coordinates.

3. Consider the collection of partial derivatives we got as the components of a tensor of rank 2 in the X-coordinates.

4. Switch back to our original coordinate system Y, and re-express in that original system the tensor we got, using the tensor equations linking X and Y.

Let's look at what we get and then comment on each term. We find that, with this new definition, the derivatives at P are the old ones corrected by the addition of another term:

$$D_r V_m = \partial_r V_m - \Gamma^t_{rm} \, V_t \tag{10}$$

Here is how to read equation (10):

1. The notation $D_r V_m$ is by definition the partial derivative of V_m with respect to the r-th direction in X obtained from the procedure, that is, in the Gaussian normal coordinates X, and then re-expressed in the Y-coordinates.

2. The term $\partial_r V_m$ is the ordinary partial derivative of V_m with respect to the r-th direction in Y calculated directly in the Y-coordinates. Note that ∂_r is a shorthand for $\frac{\partial}{\partial Y^r}$.

3. Lastly, $-\Gamma^t_{rm} \, V_t$ is the additional term due to the fact that the coordinates Y themselves vary in the vicinity of P. The minus sign is a pure convention. This whole second term on the right-hand side of equation (10) must clearly be proportional to V_t. If you double the size of V_t, it must be twice as

big. The coefficient Γ^t_{rm} in front of V_t is a new mathematical object appearing in the differentiation procedure. We shall talk more about it.

The term $\Gamma^t_{rm} V_t$ does not have a derivative related to the vector because it doesn't come from the fact that the vector is changing. It comes from the fact that the coordinates are changing in the vicinity of P.

The right-hand side of equation (10) is what you will get if you take a vector, differentiate it in Gaussian normal coordinates, and then transform the double-indexed collection of derivatives to other coordinates as a tensor. In any other coordinate system, what you will get is the usual derivative in that coordinate system *minus an object* times the components of V themselves. As usual, $\Gamma^t_{rm} V_t$ means the sum over t. Equation (10) holds in any arbitrary coordinate system. Of course, the V_m's or V_t's and the Γ^t_{rm}'s depend on the coordinate system. Notice that there is not only one number Γ^t_{rm}: they are a three-indexed collection.

We succeeded in defining a kind of derivative of a vector, which actually is a tensor. It is called the *covariant derivative* of V_m.

Andy: *That's cool! And I'm guessing it is called the covariant derivative because the index is downstairs.*

Lenny: *Brilliant Andy! But wrong. In this context the term* covariant *has nothing to do with the placement of indices. It simply means that the procedure gives back another tensor. You could raise the indices and make them contravariant, and it would be the contravariant components of the covariant derivative. Get it?*

Andy: *Huh? Don't go too fast ...*

Lenny: *Tell you what: forget covariant; let's just call it the shmovariant derivative*

Andy: *I think I get it. So equation (10) is the covariant component of the shmovariant derivative?*

Lenny: *Yeah. Now can we go back to calling it the covariant derivative?*

Christoffel Symbols

The coefficients Γ^t_{rm} have two names: *connection coefficients* and *Christoffel symbols*.

The name *connection coefficients* comes from the fact that they connect neighboring points and tell us how to calculate the rate of change of a vector field from one point to another nearby, even though the coordinate system may be changing.

They are also called *Christoffel symbols* after Elwin Christoffel.[5] They have been known on occasion as the "Christ awful" symbols because they seem complicated. With some practice, however, the reader will discover that they are not that complicated. They are just an extra linear term. But I grant you that they are complicated and unlikeable enough.

Let's investigate what follows from the definition of the covariant derivative and the Christoffel symbols. We are not going to prove every single fact we state, because there are just too many little pieces. But they are easy to check.

It follows from the definition of the covariant differentiation – namely, to differentiate a vector V at a point P, go to a set of Gaussian normal coordinates at P, differentiate the vector in the ordinary manner, treat the object you obtain as a tensor with two indices, change coordinates, etc. – that the Christoffel symbols have a symmetry:

$$\Gamma^t_{rm} = \Gamma^t_{mr} \tag{11}$$

There are generalized Riemannian geometries, also called geometries with torsion, in which this symmetry is not true. But those geometries are not widely in use in ordinary gravitational theory. The geometry of general relativity is the Minkowski–Einstein geometry, which is an extension of Riemannian geometry with a non-positive definite metric. But it doesn't involve torsion. So the Christoffel symbols we will use will be symmetric as in (11).

---

[5]Elwin Bruno Christoffel (1829–1900), German mathematician at the University of Strasbourg – when it was in the German empire between 1871 and 1918 – who did fundamental work in differential geometry.

To build our physical intuition, let's observe that calculating the derivative in Gaussian normal coordinates, which are almost flat, or as flat as can be, and then treating what we obtain as an object in its own right, is very similar to what we do in gravitational theory when we evaluate something in a free-falling frame.

For example, in lecture 1, in a free-falling frame we calculated how light moved across an elevator, and then we transformed it to the frame of reference in which the elevator was accelerating.[6] That is closely related to the operations we have been doing in this lecture: we calculate something because we know how to do it in coordinates that are as flat as possible. That would be a free-falling frame in general relativity. Then we transform it in any coordinate we like, accelerated coordinates or anything we need, and we translate the statement from one coordinate system to another. In the construction of the covariant derivative, the calculation of the variation of a vector from point to point is done first in Gaussian normal coordinates, and then it is transformed in any coordinate system. Equation (10), reproduced here

$$D_r V_m = \partial_r V_m - \Gamma^t_{rm} V_t \tag{12}$$

is the form that you get for the corresponding collection of components. It is a tensor. However, $\partial_r V_m$ is not a tensor. Therefore $\Gamma^t_{rm} V_t$ cannot be a tensor. And Γ^t_{rm} cannot be a tensor either.

We will see that the Γ^t_{rm}'s are built up out of the derivatives of the metric $\partial_r g_{mn}$. In fact in a coordinate system in which the derivatives of the metric are zero, the Christoffel symbols are zero. But a tensor, if it is zero in one coordinate system, is zero in every coordinate system – that's, among other things, what makes them so useful. So that is another way to see that they can't be tensors.

Let's look now at the covariant derivative of higher-rank tensors, because we will need this for curvature. Suppose that we have a tensor with more than one index, say

$$T_{mn}$$

---

[6]If we consider the trajectory of a light ray in the vicinity of a mass, and a small laboratory free-falling toward that mass, and the light ray crossing the laboratory, we will see, in the laboratory, the light ray going straight.

and we want to differentiate it covariantly along the r-th axis. We denote the resulting tensor

$$D_r T_{mn}$$

Its expression is the analog of equation (12), except that for every index in the tensor T_{mn}, there will be a term like $\Gamma^t_{rm} V_t$. Let's see in more detail how it works.

We start by working only on the m index, letting n be passive. Writing the equivalent of (12), we get

$$D_r T_{mn} = \partial_r T_{mn} - \Gamma^t_{rm} T_{tn} - \ldots$$

This is only a part of what we want. We have to do exactly the same with the n index, letting this time m be passive:

$$D_r T_{mn} = \partial_r T_{mn} - \Gamma^t_{rm} T_{tn} - \Gamma^t_{rn} T_{mt} \qquad (13)$$

That is the form of the covariant derivative at point P of the tensor T_{mn}. The rule is the same: we switch to Gaussian normal coordinates at P, and we do the ordinary differentiation of the tensor with respect to each direction X^r. This adds one more index to the collection of components that formed T_{mn}.[7] Then we re-express the new tensor in the original coordinate system with the usual tensor equations (equations (16) and (17) of lecture 2 and their generalizations).

This allows us to differentiate any tensor. At the moment we are only dealing with tensors with covariant indices. We will come in a moment to tensors with contravariant indices.

The reader may wonder: what is all this intricate business of covariant differentiation of tensors for?

It is for comparing things at different points. We want to be able to talk about rates of variation of things along coordinate lines, with objects that have an existence irrespective of the system of coordinates we work with.

---

[7]In order not to clutter equations, we no longer use prime signs for one system and un-prime for the other. The last time we did it in this lecture was in our comments following figure 4.

Remember that a vector in ordinary three dimensions has an existence irrespective of the basis we are using. For certain works and calculations with it – not all of them – we need a representation of the vector in a basis. The collection of components to represent it and work with it is different from one basis to another, but the vector we are talking about is the same.[8]

Where are we going to use covariant derivatives? Answer: in field equations. *Field equations* are going to be differential equations that represent how a field changes from one place to another. But we want them to be the same equations in every reference frame. We don't want to write down equations that are specific to some peculiar frame. We want them to be valid in general. That is, if they are true in one frame, they will be true in all frames. That means they have to be tensor equations. So we have to know how to differentiate tensors to get other tensors.

Another point worth stressing: the Christoffel coefficients will be present in equation (13) even in a flat space – like a plane, or this page, or the ordinary 3D Euclidean space – *if you chose funny coordinates*; see figure 1 of lecture 2. That is an important point: terms like $\Gamma^t_{rm}T_{tn}$ are there even in flat space if you are using funny coordinates. In fact if you choose any coordinates in which the derivatives of the g_{mn}'s are not zero, that is, in which the coordinates vary from point to point sinuously (viewed from an embedding space, for instance), terms like $\Gamma^t_{rm}T_{tn}$ will be present.

The presence of terms like $\Gamma^t_{rm}T_{tn}$ in the covariant derivative of a tensor is not a characteristic feature of curved spaces, it is a feature of curved coordinates.

To begin to use our new tool, let's apply equation (13) to the metric tensor itself. There is something special, however, about the metric tensor: in Gaussian normal coordinates, its derivatives are all zero. It's easy to check. But that in turn implies the following fact:

---

[8]It is even like the difference between things and their notations. When we talk about 12 in the decimal system or 1100 in the binary system, we are actually talking about the same thing: the number twelve – which some other people call douze (in French) or shí'èr (in Chinese, written in pinyin) or dvenadtsat' (in Russian, written with roman characters), etc.

The covariant derivative of the metric tensor is zero.

This simple observation turns out to be very powerful. It is what allows us to compute the Christoffel symbols. Let's write equation (13) for the metric tensor:

$$D_r g_{mn} = \partial_r g_{mn} - \Gamma^t_{rm} g_{tn} - \Gamma^t_{rn} g_{mt}$$

We know that this is zero, because, as said, the ordinary derivative of the metric tensor in Gaussian normal coordinates is zero. So, in any coordinate system, we have

$$\partial_r g_{mn} - \Gamma^t_{rm} g_{tn} - \Gamma^t_{rn} g_{mt} = 0 \qquad (14a)$$

Let's write the same equation, except with permutation of the indices. It is a little trick to get as much juice from the Christoffel symbols as we can and, eventually, via some nice cancellations, to be able to isolate one Christoffel symbol and express it in terms of the ordinary partial derivatives of g with respect to the axes in any coordinate system. Equation (14a) becomes

$$\partial_m g_{rn} - \Gamma^t_{mr} g_{tn} - \Gamma^t_{mn} g_{rt} = 0$$

The middle term, by symmetry, can be rewritten interchanging m and r:

$$\partial_m g_{rn} - \Gamma^t_{rm} g_{tn} - \Gamma^t_{mn} g_{rt} = 0 \qquad (14b)$$

Similarly we can write

$$\partial_n g_{rm} - \Gamma^t_{rn} g_{tm} - \Gamma^t_{mn} g_{rt} = 0 \qquad (14c)$$

Let's write these three interesting equations next to each other to look at them more conveniently:

$$\begin{aligned}
\partial_r g_{mn} - \Gamma^t_{rm} g_{tn} - \Gamma^t_{rn} g_{mt} &= 0 \\
\partial_m g_{rn} - \Gamma^t_{rm} g_{tn} - \Gamma^t_{mn} g_{rt} &= 0 \\
\partial_n g_{rm} - \Gamma^t_{rn} g_{tm} - \Gamma^t_{mn} g_{rt} &= 0
\end{aligned} \qquad (15)$$

How can we add them, or subtract them, or do something clever, to isolate only one of the terms with a gamma?

Let's add equation (14b) to (14c) and subtract (14a). Of course, we will get $\partial_n g_{rm} + \partial_m g_{rn} - \partial_r g_{mn}$ plus some other terms. But the middle term of (14a), $\Gamma^t_{rm} g_{tn}$, will disappear, and so will the last term of (14a), $\Gamma^t_{rn} g_{mt}$. We will be left with twice the same last term with a gamma, $\Gamma^t_{mn} g_{rt}$. So we are in luck: (14b) + (14c) − (14a) yields

$$\partial_n g_{rm} + \partial_m g_{rn} - \partial_r g_{mn} = 2\Gamma^t_{mn} g_{rt} \qquad (16)$$

We are still not done. We would like to get Γ^t_{mn} by itself. Our goal, indeed, is to find out what the Christoffel symbols are in terms of derivatives of the metric. We are almost there. The reader may have guessed what we are going to do.

Notice that equation (16) shows that if all the derivatives of the metric are zero, then the Christoffel symbols must be zero.

How are we going to get rid of the g_{rt} on the right-hand side of equation (16)? The answer comes from recalling that g_{rt} has an inverse. We saw that in the form of matrix equations, as well as in the form of tensor equations; see equation (34) of lecture 2. We multiply both sides of equation (16) by the inverse tensor, and move also the factor 2. It yields

$$\Gamma^t_{mn} = \frac{1}{2}\, g^{rt}\, [\, \partial_n g_{rm} + \partial_m g_{rn} - \partial_r g_{mn}\,] \qquad (17)$$

This is the expression of the Christoffel symbols in terms of the ordinary derivatives of the metric tensor.

It is rather simple. The indices m and n are symmetric. You can interchange them, the Christoffel symbol won't change. There are two positive terms and one negative term. It is not very complicated. The problem is that there is a boatload of them. When you think about a four-dimensional space and let all the coefficients range from 1 to 4, there is just a lot of Christoffel symbols. That is what makes doing calculations in general relativity a very tedious business. Intrinsically there is nothing hard about it. But doing a calculation in a general relativity context usually fills page after page of nothing more complicated than just computing these derivatives and assembling them together.

Equation (17) holds for any coordinate system and any metric tensor. Notice that all our calculations are *at one point* P. Whatever coordinate system our manifold is equipped with, we position ourselves at a point on it, consider the metric tensor g_{mn} there, and calculate the gammas there with equation (17). The use of Gaussian normal coordinates at P was just for intermediate reasoning, calculation, and proof purposes. We are now back in the initial coordinate system of our space. The g_{mn}, g^{mn}, and Γ^t_{mn} all depend on P; they are fields. But equation (17) is general. At every point, it expresses the connection coefficients – the other name for the Christoffel symbols – in terms of the derivatives of g. These connection coefficients enable us to figure out how any vector or tensor varies when we move a bit along a coordinate line.

The problem with the Christoffel symbols is that they are not tensors.[9] They can be zero in one frame of reference and not zero in another. For example, in a set of Gaussian normal coordinates at point P, all of the Γ^t_{mn} are equal to zero. This can be seen in many ways. Since the metric tensor in that case is constant (even equal to the Kronecker-delta tensor, but that is not necessary), equation (17) tells us that $\Gamma^t_{mn} = 0$. Yet in some other coordinate systems, the Christoffel symbols are not.

We mentioned several times that even in an intrinsically flat space, we can have coordinates such that the metric tensor is not constant. Then the Christoffel symbols won't be zero. Let's repeat: *the Christoffel symbols are related to the coordinate system, not to the intrinsic geometry of the space.*

A sphere is intrinsically non-flat. In the polar coordinates θ and ϕ (see lecture 1, figure 14), the components of g are not constant, therefore the Christoffel symbols are not zero in that system of coordinates. Even on a sphere, however, at any given point we can build a set of Gaussian normal coordinates – like maps do – then the Christoffel symbols at that point will be zero.

---

[9] At this point, after pages of higher mathematics, the reader may like to pause and remember a simple and familiar example of something very useful, yet which doesn't have the nice properties of a tensor: at a point P, an ordinary contravariant vector is a tensor (of rank 1 with one superscript index), but the first component of the vector is not a scalar tensor (a tensor of rank 0), because it will change depending on the coordinate system.

Exercise 1: Explain why the space can be flat and nevertheless the Christoffel symbols not zero.

Exercise 2: Explain why the covariant derivative of the metric tensor is always zero.

Exercise 3: On Earth, with the polar coordinates θ for latitude and ϕ for longitude, find

1. the metric tensor g_{mn}

2. its inverse g^{mn}

3. the Christoffel symbols at point (θ, ϕ).

When we meet all this for the first time, it appears conceptually tricky. But at the end of the day, the rule is simple: calculate the Christoffel symbols and, in many contexts, replace ordinary derivatives with covariant derivatives.

You could write your equations in Gaussian normal coordinates. Then they would just involve ordinary derivatives, and we would not have to wade through a river of Christoffel symbols. But if you want the same equations in general coordinates, then replace the ordinary derivatives by covariant derivatives.

That is the procedure. It will require the reader to think about it. You will have to sit down, carefully follow the reasoning, do the exercises we propose and many more. Then what we are doing will become clear.

Curvature Tensor

What is curvature? It is easiest to start with two-dimensional curvature. Intuitively it is easy to understand: it is a characteristic of something that is round and cannot be flattened out. But we are going to give it some more mathematical definition. How do we probe for curvature?

Let's begin by drawing a space that is curved. A sphere is curved, yet a curved surface that resembles a cone will be more illuminating for our purpose. It is going to be a cone with a round summit; see figure 6.

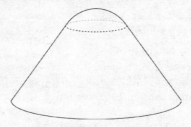

Figure 6: Cone with a rounded summit.

Think of the top of a mountain the sides of which are nice and flat like those of a volcano, and the top is round.

If you are away from the top of the mountain, below the dotted line, around you the surface is flat.[10] It may not look flat because, like the furled page in figure 10 of lecture 1, we represented the mountain embedded in 3D Euclidean space. But the surface is what mathematicians call *developable*: any section with no hole in it, cut from the side of the mountain, can be flattened onto a plane without distortion.

The rounded cone only differs from a flat space in the vicinity of the summit. To see that, just take the same space below the dotted line but continue it so that it really does form a genuine cone.

---

[10]Technically it is flat because one of its two principal one-dimensional curvatures is zero. Consequently its Gaussian curvature, which is the product of the two, is zero, and it can be shown that the surface (below the dotted line) necessarily can be unfurled into a flat one. For a good chapter on curves and surfaces, see, for instance, the book that Andy is a fan of by Alexandrov, Kolmogorov & Lavrentiev, *Mathematics*, Dover, 1999, chapter VII.

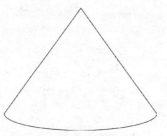

Figure 7: Genuine cone.

Then slice the cone along a generatrix, i.e., a straight line on the cone going to the top. And open it up. You can lay out flat on a plane the shape that you get. It is a disk with a missing piece, see figure 8.

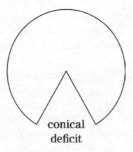

Figure 8: Cone opened up and laid flat (smaller scale and smaller angle than in figure 7).

The missing piece is called the deficit angle, or the *conical deficit*. We can see that the bigger the conical deficit is, the pointier the cone will be.

Now, *on the flat surface* of figure 8, let's consider a collection of identical vectors arranged around the shape as shown in figure 9.

On the flat surface, all the vectors point in the same direction. But when we fold the shape to form the cone, we see that the vectors no longer point in the same direction. Think of them as very small so that they don't have to be bent. The first one on the left is along a generatrix, but the last one on the right points away from it.

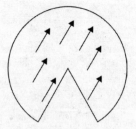

Figure 9: Identical vectors.

We can describe this effect another way. Let's suppose that our bug on the surface has a short vector – a pointer that points in some direction lying within the surface. Whenever the bug moves, it is very careful to keep the direction of the vector fixed. In three dimensions, it might do this with the aid of a gyroscope, and you can imagine a similar apparatus in two dimensions.

You might think that if the bug travels in a closed circuit, when it gets back to the starting point its vector will point in the same direction as when it started. But you'd be wrong if its orbit took it around the tip of the cone; in that case the vector will undergo a rotation. You can see that in figure 9.[11] The angle of the rotation is the same as the conical deficit.

Exactly the same is true on the rounded cone of figure 6: if we take a vector on the flat side, below the dotted line, and we carry it around the mountain in such a way that, when the surface is opened up and laid on a plane, the vector is always pointing in the same direction, by the time we get back to the other side, it will be pointing in a rotated direction.

That is the effect of curvature: when you parallel transport a vector in a closed loop around a region with curvature, the vector undergoes a rotation, despite all your efforts to keep it parallel to its initial direction.

---

[11]It is also clearly shown in figure 3 of lecture 4.

There is another way to say this, which is equivalent and actually more useful. Consider a curved space with some curvature at point P, as in figure 10. Take a vector field and differentiate it along one axis (first displacement in figure 10). Then differentiate it along the second axis (second displacement in figure 10). That is, you consider the vector field at P; then you move a bit along one axis and consider the new value of the vector field at I; then you move another bit along the second axis and consider the value of the vector field at Q.

Over each displacement, the vector will change. How will it change? The vector will change typically by differentiating it along the two axes in sequence. We first differentiate the vector along one axis and then differentiate it along the second axis. This will produce a small change in the vector due to the two derivatives.

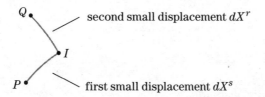

Figure 10: Displacements to differentiate a vector field along two axes.

The total change in the vector consists of two changes. And that total change is proportional to a second derivative. That is true in any coordinates: if, to compare the vectors at Q and at P, you compared the vector at I with the vector at P, and then compared the vector at Q with the vector at I, what you would be calculating is the second partial derivative of the vector with respect to the two directions.

In figure 10, if the first displacement is along the direction X^s and the second displacement along the direction X^r, then the variation of the m-th component of the vector V – let's say it has covariant indices – would be

$$D_r D_s V_m \qquad (18)$$

This expression is calculated covariantly. In Gaussian normal co-ordinates, expression (18) would just contain ordinary derivatives.

We could have also gone in the other direction, as in figure 11. That is to say, we could have gone first in the r direction and then in the s direction and calculated the way the vector changes from P to J and then from J to Q.

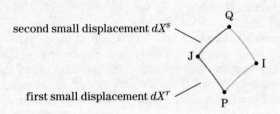

Figure 11: Displacements in the other order.

The variation of V would then be

$$D_s D_r V_m \tag{19}$$

Ordinarily, and in flat space in general, expression (18) and expression (19) are equal to each other:

$$D_r D_s V_m = D_s D_r V_m \tag{20}$$

This is just a version of the fact, in calculus, that the partial derivatives of a nicely behaved function of several variables can be taken in the order you like (see interlude 3 in volume 1 of TTM).

Equation (20) is not true in curved space. In that case the difference between the two sequences of differentiation, which is

$$D_r D_s V_m - D_s D_r V_m$$

can be thought of as taking the vector around the closed loop

$$P \to I \to Q \to J \to P$$

Let's go back to our cone, either the genuine cone (figure 7) or the cone with a rounded top but looking at the part below the dotted line (figure 6). Consider a vector field that, when the cone is opened and laid flat, is constant. Fold the flat shape to form the cone. We discovered that if we follow the vector field on a closed loop around the top, we don't get back to the same vector we started with. This is due to the following fact, which is important enough to stress:

In flat space covariant derivatives are interchangeable. In curved space they are not.

That will enable us to test whether the space is flat or not.

We will test whether differentiating tensors, and in particular vectors, in opposite order gives the same result.

- If the answer is yes everywhere in the space for any vector, then the space is flat.

- If we discover that there are places in the space where the order of differentiation gives different answers, then we know that the space has some kind of defect in it (like the point of the genuine cone) or has curvature (like the summit of the rounded cone).

All we have to do is compute the second covariant derivatives of a vector in opposite order and compare them. In principle it is not complicated. In practice it will be a little complicated, but will remain manageable. We have all the tools at our disposal. Now it is a mechanical operation, consisting of pure plug-ins. We will sketch the steps, and then give the answer.

We start with a vector expressed with covariant components:

$$V_n$$

We compute its covariant derivative in the r direction:

$$D_r V_n$$

Then we differentiate this, still covariantly, in the s direction:

$$D_s D_r V_n \tag{21}$$

After completing this step, we will interchange the indices s and r and subtract.

Let's replace the first covariant derivative of V_n, with respect to r, by its expression given in equation (12). We get

$$D_s D_r V_n = D_s \left[\partial_r V_n - \Gamma^t_{rn} V_t \right]$$

Notice that $[\partial_r V_n - \Gamma^t_{rn} V_t]$ is a tensor. We know how to differentiate it: use equation (13). Continue to crank mechanically the calculations.

In the end, the difference between the two second-order covariant derivatives yields a tensor, denoted $\mathcal{R}^{\ \ \ t}_{srn}$, multiplied by V_t:

$$D_s D_r V_n - D_r D_s V_n = \mathcal{R}^{\ \ \ t}_{srn} V_t \qquad (22)$$

Here is the tensor:

$$\mathcal{R}^{\ \ \ t}_{srn} = \partial_r \Gamma^t_{sn} - \partial_s \Gamma^t_{rn} + \Gamma^p_{sn} \Gamma^t_{pr} - \Gamma^p_{rn} \Gamma^t_{ps} \qquad (23)$$

There are two terms involving derivatives of Christoffel symbols and two terms that are sums over p of products of Christoffel symbols.

The tensor $\mathcal{R}^{\ \ \ t}_{srn}$ is the *curvature tensor*, also called the *Riemann curvature tensor* or *Riemann–Christoffel tensor*.

It has a complicated expression. It is even more complicated when you remember that the Christoffel symbols are given by the equation

$$\Gamma^t_{mn} = \frac{1}{2} \, g^{rt} \, \left[\, \partial_n g_{rm} + \partial_m g_{rn} - \partial_r g_{mn} \, \right]$$

Let's see what are the elements in the curvature tensor given by equation (23). The Christoffel symbols involve derivatives of g. So differentiating again produces second derivatives of g. Remember that the second derivatives of g are the things that we cannot generally set equal to zero. For the first derivatives of g, we saw that we can find a frame of reference where they are equal to zero. But for the second derivatives of g, we can't. So by the time we are finished calculating the curvature tensor, the second derivatives of g

have come into it. The second derivatives are testing and probing out the geometry of the surface a little more thoroughly than just the first derivatives. In a similar way, in the theory of functions, when at a point x you know $f(x)$ and $f'(x)$ and $f''(x)$, you are better off than if you just know $f(x)$ and its first derivative $f'(x)$.

Thus the curvature tensor contains second derivatives of the metric g, and it has squares or quadratic things involving first derivatives of g. It is a complicated creature. If we were to actually write it in terms of the metric, or we were to try to calculate it for a given metric, it could rapidly fill up pages. But conceptually what it is doing is simply calculating the difference in a vector if you transport it around the loop in figure 11, keeping it parallel to itself, as much as you can locally at every point, until you have come all the way around. It calculates the little change in a vector in parallel transport going around a loop.

The curvature tensor has a complicated formula, but we can calculate it. We can put the metric tensor into a computer and ask the computer: "Is the curvature tensor 0?" It is even better if you have software that can do algebra. If you have the metric in some algebraic form, you can do all the operations of equations (17) and (23) and then test out whether the curvature tensor is zero everywhere. If the curvature tensor is zero everywhere, that is, all its components are zero everywhere, then your space is flat.

We shall study the curvature tensor a little more. As said, it is a complicated thing. Its main use is to tell us whether the space is flat. And, if not, how unflat it is.

It is closely related to a quantity in gravitational physics. Can you guess which one? A local quantity that tells you that the space is not flat. It must be something telling you whether there is really a gravitational field present or not. Answer: it is the tidal forces. It is exactly related to tidal forces, those things that in a gravitational field squeeze bodies one way and stretch them another way. Tidal forces are represented by the curvature tensor.

Here is another way to get a feel for what the curvature tensor is. Imagine a surface that is flat away from a point in the center where there is a bulge. It doesn't have to be a rounded cone. It can simply be a plane with a bulge, as in figure 12.

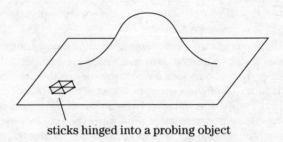

sticks hinged into a probing object

Figure 12: Probing the curvature with a little structure of Tinkertoy sticks.

You have a small structure of Tinkertoy sticks, all hinged at their extremities, so that their directions can move freely from each other, while remaining attached. At first, the probing structure lies flat, without stress or distortion, in a flat part of the surface, because the probe is itself flat.

Then you start moving the probe. While you move it in the flat region, nothing happens to it. It remains perfectly happy. It doesn't get stretched, it doesn't get distorted or deformed. This would have also been the case on the side of the rounded cone away from the summit, by the way.

What happens when you try to move the probe into the curved region? Then it simply can't follow the curvature without having to stretch or compress some of its lengths. It has to follow the metric properties of the curved space. In particular, if you go around the probe, what you are doing somehow is sampling the double covariant derivatives of equation (18) or (19). You are going to find out that various angles between sticks change from their value in flat space. The lengths of the sticks shift too; they get stressed, they get deformed. The measure of how much the probe gets stressed locally is given by the curvature tensor.

The curvature is an important property because, if you are in a region where there is curvature, you can feel it, either with tidal forces in a gravitational field or with the probe in the experiment of figure 12.

Uniform gravitational fields don't have curvature. That is why in free fall, in a perfectly uniform gravitational field, you simply feel nothing. Indeed, uniform gravitational fields don't create tidal forces. Of course, perfectly uniform gravitational fields don't really exist in nature. You can simulate one with acceleration, but you cannot see one in nature. They exist only approximately on the surface of big massive objects, if you limit yourself to a small solid angle. This leads us to a last remark.

Tidal forces, or curvature on a surface, have a bigger effect on bigger objects. The 2000-mile man in free fall toward the Earth will feel tidal forces more strongly than a free-falling bacteria. Similarly, in figure 12, if the probe is small compared to the bulge, it won't be much deformed when it goes over it. Whereas, if it were a bigger Tinkertoy structure, made for instance of many more hexagons, covering a larger region of the plane, like floor tiles, but still hinged so that any two connected sticks can change their direction from each other, but not their length, then the probe would feel the curvature more strongly.

Let's pause and see where we have arrived. At the end of this third lecture, we have reached the curvature tensor. It is complicated. Its expression is given by equation (23). I wish we could do without the sea of symbols and indices, but that is not possible if we really want to understand the nature of the curvature tensor. Nevertheless, the essential point is that we can compute it.

Often you will be presented with the metric tensor in some analytic form. There will be a formula for it. With the formula you can do differentiation. Everything will consist then of analytic functions that you can calculate.

Therefore we have finally reached our initial goal. Remember that it was to find a method to determine whether a space is flat. By definition, the space is flat if there exists a set of coordinates in which the metric tensor is everywhere equal to the Kronecker-delta tensor.

The idea of trying out every possible set of coordinates, and checking them at every point, was not a practical solution. So we found the curvature tensor. If it is zero everywhere, then we can find a set of coordinates such that the metric tensor is everywhere equal to the Kronecker-delta tensor. You just position yourself at any fixed point and start to build Euclidean coordinates, like we did when we built Gaussian normal coordinates. If our space has no curvature, these Euclidean coordinates won't be limited to a small vicinity.

In summary:

- *The space is flat if and only if the curvature tensor is everywhere equal to zero.*

- *The curvature tensor has a complicated form given by equation (23). But when we know the metric, the curvature tensor can be computed at every point of the space. Therefore it is a practical tool.*

Notice that knowing the metric of the space at every point is not a stringent condition; it is the basic knowledge we must have about it. If we don't know its metric, we really don't know what our space looks like.

We are finished with our mathematical study of Riemannian geometry, metrics, tensors, curvature, etc. The interested reader who wants to go further into the mathematical aspects of these topics can open any good manual on differential geometry oriented toward applications. As far as we are concerned, our new toolbox is now complete. We are ready to use it.

In the next lecture, we will enter into gravity land. We will see what has to change to go from Riemann geometry to Einstein geometry. Then we will study a famous simple example: the Schwarzschild geometry. It is the geometry of a black hole, a star, or any gravitating mass.

Lecture 4: Geodesics and Gravity

Andy: *Lenny, if I keep going straight ahead, following my nose, do I follow a geodesic?*

Lenny: *Yup, that's the idea. But what straight ahead means depends on the geometry of the surface. It's affected by the presence of masses.*

Andy: *So a guy on a drunk going from bar to bar, crossing the street several times, he follows a geodesic?*

Lenny: *Well I suppose you could say that the bars are like masses; they exert an attractive force.*

Introduction

In this lecture we gradually move from Riemannian geometry, where the squared distance between two points is always a positive number, to Minkowski geometry, where the squared "distance" between two points, i.e., two events in space-time, can be positive, null, or negative.

Let's begin by recalling the basic formulas established in the previous lecture on Riemannian geometry. Many of them will transfer with no change, but the metric will be less intuitive.

The covariant derivative of the simplest kind of tensor (leaving aside scalars), when we consider a *covariant vector*,[1] is given by the formula

$$D_r V_m = \partial_r V_m - \Gamma_{rm}^t V_t \tag{1}$$

where Γ_{rm}^t is called a Christoffel symbol.

For a tensor with more covariant indices, the formula is a simple generalization of equation (1), carrying an extra term with a Christoffel symbol for each index. For instance

$$D_r T_{mn} = \partial_r T_{mn} - \Gamma_{rm}^t T_{tn} - \Gamma_{rn}^t T_{mt} \tag{2}$$

Equations (1) and (2) are valid in any coordinate system. At any given point, if we are locally using a coordinate system that is as close as possible to Cartesian, then the Christoffel symbols are zero, and the right-hand sides reduce to their first terms, that is, to ordinary derivatives.

We now turn to a specific tensor: the metric tensor. Cartesian coordinates are by definition a coordinate system in which the metric does not depend on the point P, moreover is equal to the Kronecker-delta tensor. A space in which such a system can be found is called flat.

Similarly, *locally*, a Gaussian normal coordinate system is one in which the metric tensor is locally the Kronecker-delta tensor up to second order (that is, still behaving like the Kronecker tensor in the first order but not in the second). Therefore, at any given point P, in a set of Gaussian normal coordinates at that point, the *ordinary partial derivatives* of the components of the metric tensor are zero:

$$\partial_r \; g_{mn} = 0 \tag{3}$$

---

[1] We saw in the previous lectures that, for us, vectors are abstract things which have a contravariant form, i.e., a collection of contravariant components, and also a covariant form. When we talk about a covariant vector, we mean, to be more rigorous, a vector expressed with its covariant components.

This is true only in a set of Gaussian normal coordinates at the given point.

As a consequence, considering the way we have defined it, the *covariant derivative* (which is always itself a tensor) of the metric tensor *in any coordinate system*, at any point P on the surface, is equal to zero:

$$D_r \, g_{mn} = 0 \tag{4}$$

Looking again at the Christoffel symbols appearing in equations (1) and (2), we saw in many ways why they are not tensors. Unlike tensors, they can be zero in one coordinate system and not zero in another. We calculated their value, in any given coordinate system, in terms of the ordinary partial derivatives of the components of the metric tensor:

$$\Gamma^t_{mn} = \frac{1}{2} \, g^{rt} \, [\, \partial_n g_{rm} + \partial_m g_{nr} - \partial_r g_{mn} \,] \tag{5}$$

Equation (5) shows again that if the ordinary partial derivatives of the metric tensor components are zero, as is the case in a best local coordinate system, then the Christoffel symbols are zero in that coordinate system. If they were tensors, they would have to be zero in any coordinate system, but they are not.

Andy: *Lenny, how can I remember equation (5)?*

Lenny: *Same way as the Gettysburg Address. Just memorize it.*

Covariant derivatives are designed to study the rate of variation of tensors, when we move in space, in a way that is frame-independent. They are rather complicated objects if we write them out in full. Christoffel symbols are a shorthand for simplifying them.

Equation (1) was the covariant derivative of a vector with covariant components. Let's talk about the covariant derivative of a vector with *contravariant* components. We denote it

$$D_r V^m$$

As always it starts out with an ordinary partial derivative, and there is another term. The calculations are exactly the same as what we did to calculate the covariant derivative of a covariant vector. To do them, remember the following trick: there is a simple relation between the covariant form and the contravariant form of a vector. We can write

$$V^m = g^{mp}V_p$$

It is a variant of the fourth of equations (14) in lecture 2. Then take the covariant derivative of each side. Since in a best set of coordinates,[2] the covariant derivative is a standard derivative, it is easy to verify that it will satisfy the rule of differentiation for a product (see lecture 2 in volume 1 of TTM):

$$D_r V^m = (D_r g^{mp})V_p + g^{mp}(D_r V_p) \qquad (6)$$

On the right-hand side appears the covariant derivative of the inverse metric. Just like $D_r g_{mp} = 0$, see equation (4), it is easy to prove that it must also be true for the inverse metric: $D_r g^{mp} = 0$. Therefore the first term disappears, and equation (6) becomes

$$D_r V^m = g^{mp}(D_r V_p) \qquad (7)$$

We know how to calculate the covariant derivative of a vector with lower indices: it is equation (1). If you plug that in equation (7), after some algebraic manipulation you will find the formula for covariantly differentiating a vector with a contravariant index, that is, with an upper index. Here is the result:

$$D_r V^m = \partial_r V^m + \Gamma^m_{rt} V^t \qquad (8)$$

As before, the formula begins with a simple derivative. Then it has a term that would be zero in a set of best coordinates, because the covariant derivatives would simply be the ordinary ones. However they are not zero in general coordinates. In this second term with the Christoffel symbol, there is a sum over t. Generally speaking in equation (8), we check that all the indices are in place as expected. The only peculiarity is the plus sign instead of the minus sign that appeared in the covariant derivative of a vector with a lower index, as in equation (1). That minus sign was a convention. Here too, but it must be the opposite sign.

---

[2]That is the informal way we call local Gaussian coordinates.

Just as we generalized the covariant derivative of a covariant vector to tensors with covariant indices when we went from equation (1) to equation (2), we can generalize the covariant derivative to a tensor with any collection of lower and upper indices. A lower index will entail an extra term with a Christoffel symbol with a minus sign, while an upper index will entail an extra term with a Christoffel symbol with a plus sign.

We arrive at the idea of parallel transport. We already touched upon it in the previous lecture. But let's now spell it out in detail.

Parallel Transport

Suppose we have a curved surface, or a higher-dimensional curved space, and some vector field defined on it. That is, at every point of our space, there is attached a vector. In what follows, to start with, the vectors of the vector field will always be in the tangent plane – or in the higher tangent flat space – to the space.

We are interested in knowing, when we move along a curve on the space, as in figure 1, whether the field stays parallel to itself. In the figure we have represented the space and the curve, but neither the vectors of the vector field nor the curvilinear coordinates on the surface.

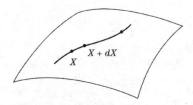

Figure 1: Vector field and curve on a space.

At each point of the curve, imagine there is a vector. Let's move along the curve. What we want to know is whether the vector (or if you prefer, the field) stays parallel to itself. "Parallel to itself" between X and $X + dX$ on the curve means the following:

The vector stays parallel to itself, when we move from X to $X +$
dX, by definition if its covariant derivative in the direction of the
curve at that point X is 0.

The covariant derivative is the difference between the vectors at
$X + dX$ and at X, as they are written in best local coordinates,
divided by the components of dX. Let's write again the tensor
that is the covariant derivative of a contravariant vector:

$$D_m V^n = \frac{\partial V^n}{\partial X^m} + \Gamma^n_{mr} V^r \tag{9}$$

Now we want to consider the derivative along the trajectory or
curve. How does the vector change from point to point? That
simply corresponds to taking the covariant derivative $D_m V^n$ and
multiplying it by dX^m. Hence, the small change in the vector is

$$D_m V^n dX^m \tag{10}$$

This formula accounts for the fact that the coordinates themselves
may evolve as we go from point to point. That is the essence of
covariant derivative.

Expression (10) is the small change in the vector V in going from
one point to its neighbor, measured by the change of its components in a set of best coordinates and then considered abstractly
in any coordinate system. Let's give it a name:

$$DV^n = D_m V^n dX^m \tag{11}$$

is the *covariant change* in the vector going from one point to a
neighboring point on the trajectory.

Let's express this covariant change with the building blocks we
have. We multiply the right-hand side of equation (9) by dX^m
and get

$$DV^n = \frac{\partial V^n}{\partial X^m} dX^m + \Gamma^n_{mr} V^r dX^m \tag{12}$$

The first term on the right-hand side has a simple interpretation.
It is the ordinary differential change in V disregarding anything
related to a possible change in coordinates. We denote it dV^n.
Equation (12) becomes

$$DV^n = dV^n + \Gamma_{mr}^n V^r dX^m \qquad (13)$$

The formula reads as follows: the covariant change in V is equal to the ordinary change in V plus a term equal to a Christoffel symbol multiplied by V^r and by dX^m. This second term is of course a double sum following the summation convention.

Equation (13) is the formula that tells you how a vector changes from point to point.

Suppose we are interested in finding a vector that is parallel to itself as we move along the curve. "Parallel to itself" means that it doesn't change as we move from X to $X + dX$. At each point X, we erect some best coordinates, and in those coordinates we test whether the vector is changing. If it doesn't change in the first order – i.e., its first derivative is zero – we say: good, the vector is constant along the little segment. We go to the next little segment, erect best coordinates at the new point, and test again. We do that all along the curve. If the sequence of tests say that the vector never changes in the first order, the vector is said to be parallel to itself along the curve.

In summary, if all along the curve the vector V satisfies

$$dV^n + \Gamma_{mr}^n V^r dX^m = 0 \qquad (14)$$

then the vector maintains a relationship of being parallel to itself.

Taking a vector from one point and transporting it like this along a given curve, in such a way that it stays parallel to itself, is called *parallel transport*. Making up a benign neologism, we say that we "parallel-transport" the vector.

A very important fact about parallel transport on a curved space is that *it is trajectory-dependent*. On the surface in figure 2, if we start at point A, take a vector V there, which lives in the tangent plane, and parallel-transport it to B, then the vector we end up with at B will depend on the path we followed from A to B.

In figure 2, we represent the vector V at A and suggested its evolution along two paths. We did not represent any coordinate system. Indeed, it is important to understand that parallel transport is dependent on the trajectory, but is independent of any coordinate system used to locate points on the surface. At each point, anyway, we use a set of best local coordinates to do the infinitesimal parallel transport of the vector there. When we arrive at B, the final vector we end up with depends not only on V of course, but also on the path we followed. The final vector depends on the bumps and troughs we encountered along the path, that is on the local curvatures along the path. Even if we came back to the same point A, depending on the loop we followed, we would end up with one or another vector. If there exists a flat *connected* region – i.e., flat and with no hole – and we follow a loop entirely in that region we will end up with the same vector V.

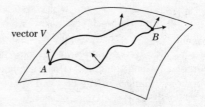

Figure 2: Parallel-transporting V from A to B. Depending on the path followed, the end vector at B is not the same.

We already saw this phenomenon on the cone – pointy or rounded, it doesn't matter – in the previous lecture. When we started with a vector on the side of the cone and parallel-transported it *around the cone*, we did not end up with the same vector. An alternative path would be not to go around the top of the cone, in which case we would end up with the same vector. This illustrates that two paths don't always lead to the same result; see figure 3.

Remember that the side of a cone is flat according to our definition, even though we see it embedded in 3D and in ordinary language it is not flat. The side of a cone is *intrinsically* flat, because any section of it with no hole can be laid out on a plane without exerting any distortion on it. More mathematically, any connected section of the side is flat because there exists a coordinate system the metric of which is the Kronecker-delta tensor over the whole section.

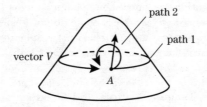

Figure 3: Parallel-transport of the vector V on a cone along two different paths, both starting and ending at A.

Parallel-transporting a vector, that is, moving it on the surface while making sure that its covariant derivative remains null, also preserves its length. It can be shown as a consequence of equation (14).

The next topic will concern tangent vectors to a curve, and whether the tangent vector stays constant or not. When the tangent vector stays parallel to itself, we will see that the curve is a *geodesic*. Geodesics are intuitive. However they are a bit trickier than we may think. For instance, on the cone of figure 4, if we go from A to B around the cone as shown (in 3D, staying parallel to the horizontal plane), we don't follow a geodesic.

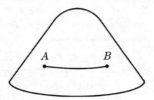

Figure 4: Going from A to B on a cone.

We will see that geodesics are shortest curves. In figure 4, going from A to B as shown is not a shortest curve. It becomes evident if we think of a very flat cone.

On Earth, boats and airplanes try to follow geodesics because they are the most economic routes. They are the so-called *great circles*. When, while we are sitting on an airplane, going to a very distant destination, the crew shows on a screen our trajectory, we are often surprised to discover that we do not follow a "straight" line. That is because a great circle is not mapped into a straight line on the usual flat representations of the Earth.

Tangent Vectors and Geodesics

We arrive at the notion of tangent vector to a curve and of geodesic. On a surface where we consider two points A and B, a geodesic between A and B is a curve with certain properties. It can be defined in several ways:

1. The curve with the shortest distance between A and B is a geodesic.

2. A curve whose length is stationary when you wiggle it is a geodesic.

3. A third, better definition looks at what happens locally along the curve: a curve that at each point is as straight as possible is a geodesic.

Of course, this last definition is more intuitive than mathematical. Let's make it more precise. If at each point along the curve, the covariant derivative of the tangent vector[3] is zero, that is, if the tangent vector doesn't change, then the curve is as straight as possible.

Let's try to build more intuition about geodesics before turning to the mathematics. Imagine a curved terrain, as in figure 5. For convenience, it is a two-dimensional example, but there is nothing special about two-dimensional spaces in defining the notion of geodesic. Secondly, imagine that we are driving a car on this terrain. And assume that the size of the car, in particular the distance between the front wheels, is small by comparison with

---

[3]When we talk about the tangent vector without further specification, it is of length one.

any curvature and that the steering wheel is locked in the straight-ahead position. We start from A in some direction, and driving straight in the above sense – never turning the steering wheel – we end up at B. Our trajectory will wind between the hills. We may also start from the top of a hill, that is, from a point with clear curvature, that doesn't change anything. The curve that we will execute with our car in the space, keeping the steering wheel straight, will nevertheless be as straight as possible. It will be a geodesic in the space.

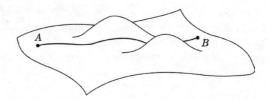

Figure 5: Driving a car straight ahead on a curved terrain.

Another way to characterize a geodesic is to say that the tangent vector along the curve is constant. We have an intuitive perception of what the tangent vector is. But let's define it more precisely. Consider a curve, and a point at coordinate X on it. And take a neighboring point; see figure 6. The points X and $X + dX$ are separated by dX, which we can also denote, in tensor style, dX^m. Consider a vector the origin of which is at X, going through $X + dX$, and of length one. Then take the limit when the second point $X + dX$ approaches the first point X. The resulting vector is called the *tangent vector* to the curve at X.

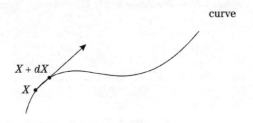

Figure 6: Construction of the tangent vector at a point.

Consider the distance dS between the two points X and $X + dX$. As we remember, it is defined by

$$dS^2 = g_{mn} \, dX^m dX^n \tag{15}$$

The way we construct the tangent vector in the X coordinate system is very simple. The m-th component of that vector is

$$t^m = \frac{dX^m}{dS} \tag{16}$$

It can be proved that equation (16) produces a vector of length one. The exercise is left to the reader. There is one such vector at each point along the curve. That, as said, is what we call the tangent vector. It points in the direction between two neighboring points and is of length one.

Let's turn our attention to curves the tangent vector of which is constant. If we plug in the tangent vector in equation (14), these curves satisfy the following equation:

$$dt^n + \Gamma_{mr}^n t^r dX^m = 0 \tag{17}$$

Equation (17) holds because once you have set your steering wheel straight ahead, you are moving in as straight a line as you can. So the covariant change of the tangent vector is zero. Next we give an example to build our intuition.

Example of Calculations with Christoffel Symbols

Building a correct intuition about geodesics is important because it is easy to be misled. This is particularly true when the surface has curvature – like a round hill. Indeed, the embedding 3D space in which we ordinarily view the intrinsic curvature of the surface suggests that the tangent vector changes when in fact it doesn't.

Consider a point P on the surface of a sphere; see figure 7. Mathematicians call such a surface a 2-sphere, because its points can be located with two coordinates. Let's use the ordinary latitude θ and longitude ϕ, and the ordinary distance we are familiar with, for instance on Earth.

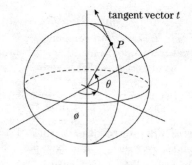

Figure 7: 2-sphere with polar coordinates.

The objective of the exercise is to show that a meridian is a geodesic. In other words, *when we follow a meridian, the tangent vector doesn't change.*

Exercise 1: We are on a 2-sphere of radius one with polar coordinates θ and ϕ, as in figure 7.

1. Show that the metric tensor of the ordinary distance is

$$\begin{pmatrix} 1 & 0 \\ 0 & \cos^2\theta \end{pmatrix}$$

2. Express the eight Christoffel symbols using this metric. Show that

$$\Gamma^1_{22} = \sin\theta \ \cos\theta$$
$$\Gamma^2_{12} = \Gamma^2_{21} = -\tan\theta$$

   and all the others are zero.

3. Show that the tangent vector to a meridian has every-where components $t^1 = 1$ and $t^2 = 0$.

4. Show that the tensor that is the covariant derivative of this tangent vector is

$$\begin{pmatrix} 0 & 0 \\ 0 & -\tan\theta \end{pmatrix}$$

5. Show that if we follow a meridian, the covariant change of the tangent vector is always zero.

Doing this exercise will show you that the actual calculations with Christoffel symbols, even on a simple example, quickly fill pages. It will also show you that even on a surface with curvature there are paths where the tangent vector doesn't change. *These are the geodesics.*

In the exercise we looked at a meridian, because the polar coordinates make it simple to study, but by symmetry any great circle is a geodesic.

In figure 7, we might feel that the tangent vector changes when we move along a meridian, *but that is because we look at the 2-sphere embedded in 3D Euclidean space.*

If we turned our steering wheel, however, and in the tangent plane swerved from our straight path, that would be another story. Then the tangent vector of our trajectory would change.

More on Geodesics

We can write equation (17) of a geodesic in a slightly neater form. Let's divide both sides of the equation by dS, that is, by the little distance between two neighboring points with coordinates X and $X + dX$; see figure 6.

Equation (17) becomes

$$\frac{dt^n}{dS} = -\Gamma^n_{mr}\, t^r\, \frac{dX^m}{dS}$$

But dX^m/dS is t^m, so we can rewrite equation (17) as

$$\frac{dt^n}{dS} = -\Gamma^n_{mr} \, t^r \, t^m \tag{18}$$

This equation only involves the tangent vector. Of course, it also involves the Christoffel symbols, but let's suppose they are given. Then equation (18) is the "equation of motion" of a geodesic.

One more thing: since the tangent vector t itself is a derivative, we can write the left-hand side as a second derivative:

$$\frac{d^2X^n}{dS^2} = -\Gamma^n_{mr} \, t^r \, t^m \tag{19}$$

Does this look familiar? If we were to think of S as some measure of time as we moved along the curve, then, on the left, the second derivative of position would be acceleration. Thus if S were like time, or were increasing uniformly with time, equation (19) would read like this: an acceleration is equal to something that depends on the metric and on the components of the tangent vector. We might even see on the right a kind of force.

For the time being, let's just observe that equation (19) has the look of a Newton equation: acceleration is equal to something that depends on the gravitational field, because, as we will see, the metric *is* the gravitational field. We will see that equation (19) replaces Newton's equation for the motion of a particle in a gravitational field. In other words, in some sense a particle in a gravitational field moves along the straightest possible trajectory. But it moves along the straightest possible trajectory not just through space but *through space-time*.

Space-Time

So far we have been studying the mathematics of curved spaces, as Riemann would have understood it. Yet Riemann's spaces were, so to speak, ordinary curved spaces in which distance was governed locally by Pythagoras theorem (in an appropriate reference frame). In particular, in Riemann's spaces the square of the distance is always positive. General relativity, however, is not just about space; it is a theory of *space-time* geometry.

The coordinates of space-time are the coordinates of space, x, y, z, and time t. Frequently we will use the more symmetrical notation:

$$x = X^1, \ y = X^2, \ z = X^3, \ ct = X^0$$

where c is the speed of light.

Space-time also has a natural measure of distance along curves or between points, called *events* in space-time, be they neighboring points or not. As in Riemannian geometry, in Minkowski geometry the distance is generally expressed through its square. But in Minkowski space this square can be zero for distinct events, or even be negative.

Let's begin with flat space-time in the analog of Cartesian coordinates. The theory of flat space-time is of course special relativity – the subject of volume 3 of TTM. You will recall that in special relativity the squared space-time distance is given by [4]

$$(\Delta\tau)^2 = (\Delta t)^2 - (\Delta X)^2$$

where $(\Delta X)^2$ is shorthand for $(\Delta x)^2 + (\Delta y)^2 + (\Delta z)^2$. We may also use the more relativistic notation just mentioned

$$(\Delta\tau)^2 = (\Delta X^0)^2 - (\Delta X)^2$$

There are three possibilities: $(\Delta\tau)^2$ can be positive, in which case the separation between the two points is said to be *time-like*; it can be negative, in which case the separation is *space-like*; or it can be zero, in which case the separation is *light-like*.

When the interval is time-like, we refer to $\Delta\tau$ as the *proper time* between the points. When $(\Delta\tau)^2$ is negative, we redefine things and call $\sqrt{(\Delta X)^2 - (\Delta t)^2}$ the *proper distance* between the points.

---

[4] For simplicity we are using units in which the speed of light is equal to 1. The more general formula is $(\Delta\tau)^2 = (\Delta t)^2 - \frac{1}{c^2}(\Delta X)^2$.

That leaves the case $\Delta\tau = 0$. Such an interval is called *null*. It represents two events that can be joined by a light ray in space-time. A null interval is also called light-like.

In flat space these definitions apply to any pair of events, far away, close, even infinitesimally close. When the events are infinitesimally close, the square of the proper distance is rewritten

$$(d\tau)^2 = (dX^0)^2 - (dX)^2$$

We have put parentheses around all the intervals to make it clear that we take the square of the interval, but in later equations we will drop these parentheses when they are not necessary, assuming the meaning of the notation is clear. Thus the previous equation will simply be written

$$d\tau^2 = (dX^0)^2 - dX^2 \qquad (20)$$

Remember too that X^0 is the same thing as t.

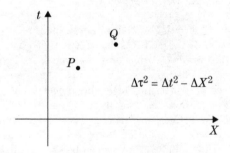

Figure 8: Proper time $\Delta\tau$ between two events P and Q. (When we write $\Delta\tau^2$, we mean $(\Delta\tau)^2$; same comment for the other intervals.)

When $d\tau^2$ is negative, it is conventional to rewrite equation (20) as

$$dt^2 - dX^2 = -dS^2$$

where S is called the *proper distance*. (The speed of light squared is in front of dt^2. But we took it to be equal to 1.)

Another convention is to write equation (20) as

$$dS^2 = g_{\mu\nu} \, dX^\mu \, dX^\nu \qquad (21a)$$

or

$$d\tau^2 = -g_{\mu\nu} \, dX^\mu \, dX^\nu \qquad (21b)$$

where readers who have read volume 3 on special relativity are familiar with the notation X^μ with a Greek upper index:

$$X^\mu = \begin{pmatrix} t \\ x \\ y \\ z \end{pmatrix}$$

According to standard convention, this is also sometimes noted

$$X^\mu = \begin{pmatrix} X^0 \\ X^1 \\ X^2 \\ X^3 \end{pmatrix}$$

where the Greek index μ runs over 0 to 3.

When we use a Latin index, we mean only the three spatial coordinates, that is, if you read X^i, this means that i runs over 1, 2, and 3. In other words, X^i runs only over the spatial coordinates.

Let's comment on equation (21a). The indices μ and ν run over 0 to 3. The equation has exactly the same form as the usual equation for the distance in Riemannian geometry that we have already often used; see equation (1) of lecture 3, for instance.

The only new thing in the Minkowski geometry is the metric tensor $g_{\mu\nu}$ or its corresponding matrix. It is still diagonal, but it has a minus 1 corresponding to the time axis, and three plus 1's for the space axes. As it plays a central role, it has a name. We use the Greek letter η (pronounced "eta") to name this matrix. And we write it $\eta_{\mu\nu}$ (pronounced "eta mu nu"):

$$\eta_{\mu\nu} = \begin{pmatrix} -1 & 0 & 0 & 0 \\ 0 & 1 & 0 & 0 \\ 0 & 0 & 1 & 0 \\ 0 & 0 & 0 & 1 \end{pmatrix} \qquad (22)$$

With this form for the metric tensor, we can check that equation (21b) expressing the proper time is the same equation as

$$d\tau^2 = dt^2 - dx^2 - dy^2 - dz^2$$

Thus far we have been speaking about special relativity. In general relativity, the metric tensor becomes a function of space and time. We then call it $g_{\mu\nu}(X)$ (where X stands for an event in space-time, i.e., a point with *four* coordinates). Equation (21b) becomes

$$d\tau^2 = -g_{\mu\nu}(X)\ dX^\mu\ dX^\nu \tag{23}$$

There is one more important thing, which we must stress at the outset, concerning the metric tensor in relativity. What is the difference between the matrix of equation (22) and the identity matrix of Euclidean metric? Well, it has a minus 1 in first position. But more importantly, there is an invariant concept about $g_{\mu\nu}(X)$: it has one negative eigenvalue and three positive eigenvalues. These four signs define its so-called *signature*. In general relativity, no matter how curved the space-time, or otherwise unfamiliar, the signature of the Minkowski metric will always be the same: three pluses and one minus, or denoted more compactly, $(-+++)$.

We are not going to spend much time dealing with this mathematical notion. Fortunately the equations of general relativity automatically guarantee that the signature is always $(-+++)$.

What does it mean that there is one negative eigenvalue and three positive? It means that there is one dimension of time and three dimensions of space. We could write a metric with two minus signs on the diagonal. It would correspond to a crazy space with two time dimensions and two space dimensions. Fortunately that is not only disallowed, but it can never occur if the equations of general relativity are correctly solved.

Other than that, all that we have done in Riemannian geometry, all the equations involving metrics, covariant derivatives, curvature, geodesics, etc. will be exactly the same in the Minkowski–Einstein space-time geometry of general relativity.

Now comes a big question: *What does flat mean in space-time?* It no longer means that there is a coordinate system in which the metric is the Kronecker-delta. It now means there is a coordinate system in which the metric has the form $\eta_{\mu\nu}$ of equation (22).

In Riemannian geometry, global flatness required the existence of a metric built with the Kronecker symbol everywhere. Similarly, in space-time, global flatness requires the existence of a system of coordinates in which the metric has the form $\eta_{\mu\nu}$ everywhere.

How do we check whether the space-time is curved? We proceed exactly analogously as we did in Riemannian geometry.

Here is a recap of the analogies that we have already made so far, as well as those we shall see:

Flat spaces
     Euclidean geometry → Minkowski geometry
     Kronecker δ tensor → η tensor
     Newtonian physics → special relativity

Non-flat spaces (always locally flat)
     Curved metric → gravitational field
     Riemannian geometry → Einstein general relativity

Before going into a space whose curvature is due to real gravitational fields, i.e., to the presence of massive bodies, we shall spend some time with a "flat" space in Minkowski geometry.

We will wind up looking at it in polar coordinates – not ordinary polar coordinates but hyperbolic polar coordinates. The name is awe-inspiring, but the concept is simple and well adapted to space-time and particles moving in it, notably particles accelerating in it.

Since we know from lecture 1 that there is a link between gravity and acceleration, and our ultimate goal is to describe relativistic motion of particles in gravitational fields, it is natural to start with studying particles accelerating in the framework of special relativity.

Special Relativity

We are in the space-time of special relativity, which we call a *Minkowski space*. Its metric is defined by the tensor

$$\eta_{\mu\nu} = \begin{pmatrix} -1 & 0 & 0 & 0 \\ 0 & 1 & 0 & 0 \\ 0 & 0 & 1 & 0 \\ 0 & 0 & 0 & 1 \end{pmatrix} \tag{24}$$

Our objective is to define the notion of a uniformly accelerated reference frame in special relativity.

We have already met a uniformly accelerated frame in lecture 1 when we illustrated the principle of equivalence with a uniformly accelerated elevator. The gravitational field of the Earth (in a small region where it can be viewed as uniform) and the apparent field we experienced in the elevator being uniformly accelerated were indistinguishable. But in lecture 1 we used Newtonian physics.

In special relativity, there is a difficulty with the notion of a uniformly accelerated reference frame. It is no longer as simple and intuitive as in Newtonian mechanics.

Let's see what is the difficulty, and how we deal with it. Consider a bunch of point-like observers, separated by a fixed distance, as in figure 9. Think of them as forming a frame.

•　　•　　•　　•　　•　　•　　•

Figure 9: Points in space with a fixed separation. We want to accelerate them "uniformly."

Suppose that the observers are accelerating along the X-axis, each having the same constant acceleration. We would think that they would remain the same distance apart. That is true in Newtonian mechanics, but in special relativity distances, time and simultaneity behave oddly as the velocity grows.

If we gave all the observers the same acceleration, we would discover that, viewed from the rest frame of the first observer, the distance to the second observer would grow. If there were strings between the observers, as they started simultaneously moving, those strings would stretch and eventually break. That is not what we would think of a uniformly accelerated reference frame, as we are accustomed to from non-relativistic physics. What is nice in non-relativistic physics, about a uniformly accelerated reference frame, is that it keeps the same structure, the same shape. The distances between points stay the same. If you had strings connecting the points, they wouldn't get stretched. But that is not so in relativity.

There is a second difficulty about the simplistic idea of uniform acceleration in special relativity:

In the naive conception of a uniformly accelerated reference frame, if we waited long enough, the observers would eventually exceed the speed of light. However in the theory of relativity, particles that we can observe never exceed the speed of light.

Uniform acceleration, to the extent that it exists and makes good physical sense, is not as simple as just moving the points in figure 9 all with the same acceleration.

We are going to construct what a relativist (i.e., a specialist in relativity theory) would call a uniformly accelerated reference frame. To do so, it will be helpful to go back to Euclidean space in polar coordinates, as in figure 10.

Surprisingly enough, the uniformly accelerated coordinate system in relativistic space-time is the analog of polar coordinates in ordinary space.

Here are some equations, expressing the coordinate transformation from polar to Cartesian coordinates, with which the reader should be familiar:

$$x = r \cos \theta$$
$$y = r \sin \theta \tag{25}$$

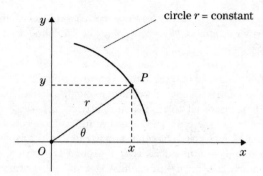

Figure 10: Euclidean and polar coordinates in the plane.

We also have

$$\cos^2\theta + \sin^2\theta = 1 \qquad (26)$$

which is the same as saying that

$$x^2 + y^2 = r^2 \qquad (27)$$

Finally there are two more equations to remember

$$\cos\theta = \frac{e^{i\theta} + e^{-i\theta}}{2}$$
$$\sin\theta = \frac{e^{i\theta} - e^{-i\theta}}{2i} \qquad (28)$$

You can check that $\cos^2\theta$ plus $\sin^2\theta$ is equal to 1. It is a simple identity, true for all possible θ. Equations (25) to (28) are the basic equations governing ordinary polar coordinates.

What is the equation of a circle around the origin? It is just

$$r = \text{constant}$$

Imagine a point moving around the circle with uniform velocity, therefore uniform angular velocity. Then the magnitude of the acceleration of that point is constant around the circle, the vector acceleration constantly pointing toward the center of the circle.

What does it have to do with relativity? We will see that in relativity we write almost the same equations to define a uniformly accelerated point.

We turn to the basic diagrammatic representation of space-time, which is the analog in special relativity of figure 10 in Newtonian physics. It is figure 11.

In figure 11, we see the light cone: the two diagonal straight lines. From volume 3, we know that they represent the trajectory of a light ray starting at time 0 from the origin and going either to the right or to the left. Remember that in the simplest Minkowski diagram, shown below, there is only one spatial dimension. Everything moves on the straight X-axis in space.

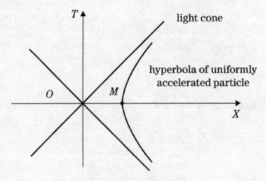

Figure 11: Light cone. It would actually be a cone if we had two spatial coordinates. Pay attention to the fact that in this familiar Minkowski diagram there is only one spatial dimension.

Notations: We use the variables X and T because later we will make a change of coordinates and arrive at variables y and t (little t) when we study a uniformly accelerated frame like the elevator of lecture 1. We will discover that uniformly accelerated frames produce a fictitious gravity, like we already observed in lecture 1.

Let's consider for a moment the analog of the circle in figure 10, but in Minkowski space. The circle of course is the locus of points a fixed distance from the origin:

$$x^2 + y^2 = r^2$$

By analogy we can consider the locus of all points a fixed space-like Minkowski distance from the origin in figure 11. It has the form of a hyperbola:

$$X^2 - T^2 = r^2$$

This suggests the following definition of a uniformly accelerated particle in special relativity.

For the moment, we will *define* a uniformly accelerated observer as one moving on a hyperbola as shown in figure 11. It is clearly not moving with constant velocity along the X-axis but is accelerating. Indeed, constant velocity would correspond to a straight line with a slope higher than 45° because things don't go faster than the speed of light.

In figure 11, we see that from the past until time 0 (i.e., the lower part of the diagram beneath the X-axis), the point or particle or observer (henceforth we shall talk of observers, rather than particles, in these positions), moves, spatially on the X-axis, to the left to a minimum point M. At M its velocity has come to zero. Therefore, in the (X, T) diagram, at M the tangent to the trajectory is vertical. After point M, the observer changes course going again to the right. In the Minkowski diagram, as the point moves up and up, the tangent to the trajectory gets closer and closer to 45°, that is, the observer moves on the X-axis closer and closer to the speed of light, without ever exceeding it.

Equations (25) describing a circle centered at O have an analog for the hyperbola in figure 11. The equations are obtained by simply replacing the trigonometric functions sin and cos by their hyperbolic counterparts, sinh[5] and cosh. The correspondence is

$$\cos\theta \;\rightarrow\; \cosh\omega$$
$$\sin\theta \;\rightarrow\; \sinh\omega$$

The mathematical definitions of the hyperbolic sine and cosine functions are very similar to those of ordinary sine and cosine. But, unlike in equations (28), there is no more $i = \sqrt{-1}$ coefficient in the exponents and the denominator of the sine:

---

[5]The hyperbolic sine is denoted sinh. Lenny likes to pronounce it "cinch" presumably because after a while we find this function very easy.

$$\cosh\omega = \frac{e^{\omega} + e^{-\omega}}{2}$$

$$\sinh\omega = \frac{e^{\omega} - e^{-\omega}}{2} \tag{29}$$

Analogously to equation (26), the reader can verify that

$$\cosh^2\omega - \sinh^2\omega = 1 \tag{30}$$

The coordinates of a point P in the (X, T) diagram are now

$$X = r\cosh\omega$$

$$T = r\sinh\omega \tag{31}$$

Equations (31) *define* r and ω from X and T. The parameter ω is not a geometric angle. But when we move along a hyperbola with the light ray trajectories as asymptotes (figure 11), it is what increases from $-\infty$ to $+\infty$, just like θ was the parameter that changed as we moved along a circle centered at the origin. On such a hyperbola, r doesn't change. The parameter ω plays on the hyperbola the role of the angle on the circle. It is sometimes called the *hyperbolic angle*.

As before, equations (31) express nothing more than a coordinate transformation between the Minkowski coordinates (X, T) and the hyperbolic coordinates (r, ω). Recall that an event in space-time corresponds to *one point* on the page. It can be located by its Minkowski coordinates (X, T) or by its hyperbolic coordinates (r, ω), or by any other system we like. That's what frames of reference are: the mathematics changes, but the physics (i.e., the space-time and what happens in it) doesn't.

In figure 12, all the points on the hyperbola have the same r. It is called the *hyperbolic radius*. Its value is the distance between O and M. It characterizes the curve. On the other hand, the hyperbolic angle ω increases up to infinity as we move on the hyperbola closer and closer to its asymptote, that is, as the observer moves spatially farther and farther away to the right on the X-axis.

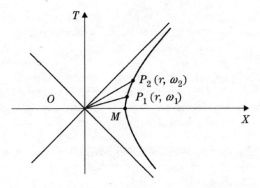

Figure 12: Hyperbola and hyperbolic coordinates.

Thus we have in Minkowski geometry the analog of the circle in Euclidean geometry: the hyperbola, as in figures 11 and 12, in hyperbolic polar coordinates given by equations (31), corresponds to a constant value r and to the parameter ω going from $-\infty$ to $+\infty$. This will be handy to study a uniformly accelerated particle because by definition it moves along such a trajectory.

The analog of equation (27) on a circle becomes, on a hyperbola,

$$X^2 - T^2 = r^2 \qquad (32)$$

Of the two coordinates r and ω, one of them is space-like, the other time-like. You can probably guess which is which, but let's go through the reasoning. On the X-axis, $\cosh \omega = 1$, so if we move to the right on this axis, we just increase r. Therefore r is like a space coordinate.

On the other hand, if from point M we travel upward on the hyperbola of figure 12, r stays fixed and we increase ω. Going upward is the analogy with traveling around the circle in figure 10 with increasing angle, but in this case it moves us in a time-like direction. So ω is like a time coordinate.

On the hyperbola in figure 12, ω is proportional to the *proper time* measured along the trajectory. More precisely it is the proper time τ measured in units of r:

$$\omega = \frac{\tau}{r}$$

If the observer who is at a fixed value of r carried a wristwatch, it would register a proper time $r\omega$ along the trajectory.

Just as there was a uniformity to the circle – at any point you could define the radius r and it was constant – there is an analog uniformity to the hyperbola: the hyperbolic radius r is constant on a hyperbola. Figure 13 shows hyperbolas for different values of r.

Let's return to the family of accelerated observers, separated by the same distance (as shown at first in figure 9). They are represented in figure 13.

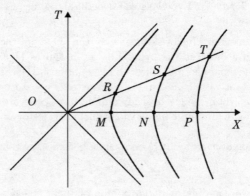

Figure 13: Hyperbolas for different values of r.

At fixed values of ω, the distances between the observers corresponding to $r = 1$, $r = 2$, $r = 3$, etc., are always the same. If in figure 13 we define the proper distance between two points M and N to be $|MN|$, then by construction

$$|MN| = |NP|$$

But it is not hard to show that the equal spacing of observers is also true at a later value of ω:

$$|MN| = |NP| = |RS| = |ST|$$

This can be checked with the tools we learned in volume 3 on special relativity.

Exercise 2: In figure 13, what is the speed, relative to the stationary frame, of the observer who sees R, S, and T as simultaneous events?

Uniform Acceleration

All of what we just saw means that as the Lorentz frame of reference accelerates, the distance between neighboring observers *does* in fact stay the same. However, there is a price to be paid. What is different from a non-relativistic accelerated frame of reference, is that the accelerations along the different trajectories corresponding to $r = 1$, $r = 2$, $r = 3$, etc. *are different.*

One can see this intuitively by looking at the various trajectories in figure 14.

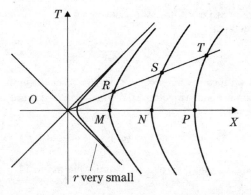

Figure 14: The hyperbola with a very small r corresponds to a very high acceleration.

On the hyperbola with a very small r, the trajectory makes a sudden change of direction when it comes close to the origin, and then speeds off to the right again very fast. That indicates that the trajectory has a large acceleration. By contrast, the trajectories farther out to the right have a much gentler change of direction, indicating a smaller acceleration.

This was all intuitive, but so far we have not defined relativistic acceleration in any systematic way.

To do so, let's first begin with velocity.

- The *ordinary velocity* of a particle is a 3-vector defined by the time-derivative of its spatial position. This is expressed by

$$v^i = \frac{dx^i}{dt}$$

- The *relativistic velocity* is a 4-vector defined by

$$u^\mu = \frac{dX^\mu}{d\tau}$$

where τ is the proper time along the trajectory.

As long as the particle is moving slowly, the space components of the relativistic velocity are very close to the ordinary velocities v.

Similarly for acceleration, the relativistic acceleration is also a 4-vector defined by

$$a^\mu = \frac{d^2 X^\mu}{d\tau^2}$$

What do we mean when we say that the acceleration is constant along a hyperbolic trajectory? We mean that the proper length of a^μ is constant. In other words,

$$|a|^2 = (a^1)^2 - (a^0)^2 = \text{constant}$$

Recall that X^0 and X^1 refer to T and X in figure 14.

Let's check it for a motion along the hyperbolic trajectory with $r = 1$. Using equations (31) and the fact that ω and the proper time τ are the same for $r = 1$, the trajectory is described by

$$X = \cosh \tau$$

$$T = \sinh \tau$$

Now it's very easy to compute the components of the acceleration. Using the properties of sinh and cosh, from these formulas one easily finds that

$$a^1 = \cosh \tau$$
$$a^0 = \sinh \tau$$

Finally using the identity $\cosh^2 \tau - \sinh^2 \tau = 1$, we get

$$|a|^2 = 1$$

Thus we indeed find that the magnitude of the acceleration is constant on the hyperbola corresponding to $r = 1$. The same is true (with a different acceleration) on each hyperbola of figure 14.

We leave it as an exercise to prove it. More precisely, for an arbitrary r, the magnitude of the acceleration is

$$|a| = \frac{1}{r} \tag{33}$$

Let's come to questions of units, as we have already done in volume 3 of TTM. Equation (33) does not look consistent unit-wise. What is the unit of acceleration? It is length divided by time divided by time, i.e., $[L]/[T]^2$. Let's rewrite this dimension as

$$\frac{1}{[L]} \frac{[L]^2}{[T]^2}$$

It is clear that to restore the units in equation (33), all we need to do is introduce a factor c^2:

$$|a| = \frac{c^2}{R} \tag{34}$$

This means that for a fixed radius R, at human scale – say, $R = 1$ meter – the acceleration of a particle (or an observer on the corresponding hyperbola) in figure 9 is extremely strong. We have to go to a very large R before we get to trajectories with a moderate acceleration.

By the way, the acceleration on a given trajectory in figure 14, for example the acceleration at point N on the hyperbola $R = 2$, is the ordinary acceleration. And it is the constant acceleration we would experience all along the trajectory.

Uniform Gravitational Field

We have introduced a somewhat arbitrary set of coordinates X, T for our stationary frame. In those coordinates, we are now going to write the equation of motion along a geodesic in the so-called accelerated coordinates r, ω. We will see that the equations look very much like a particle falling in a uniform gravitational field.

Let's first talk about the metric of the Euclidean plane in ordinary polar coordinates as shown in figure 10:

$$dS^2 = r^2 d\theta^2 + dr^2 \tag{35}$$

The matrix for this two-dimensional metric has the form

$$g_{mn} = \begin{pmatrix} r^2 & 0 \\ 0 & 1 \end{pmatrix} \tag{36}$$

Why is it not the Kronecker-delta? It is not because the space is curved, but because the coordinates are curvilinear. The space itself is flat. Indeed, it is the plane, and we can go back to Cartesian coordinates $(x,\ y)$ in which the metric is the Kronecker-delta.

Staying in the flat plane, the analog with the hyperbolic coordinates $(r,\ \omega)$ is

$$d\tau^2 = r^2 d\omega^2 - dr^2 \tag{37}$$

We are still considering only two dimensions, the time T and one spatial coordinate X. For the moment we ignore Y and Z.

The particle of interest is falling in a gravitational field along an axis denoted X (in the usual Minkowski diagram it is horizontal, but we are used to it). The coordinates Y and Z would be the other spatial coordinates. But they don't matter for the problem we are discussing. The two coordinates that we will be interested in are ω and r. Equation (37) is the metric.

Note: This might be a good place to go back and review what we did in lecture 1 when we looked at Newton's equation in the frame of a uniformly accelerated elevator. We are going to do something similar, but in Minkowski space.

Recall from the previous section that the formula for the acceleration is

$$\frac{c^2}{R}$$

To be on familiar ground, we want to set this equal to g, the acceleration on the surface of the Earth, approximately 9.80 meters per second squared, but let's use 10. It gives

$$R = \frac{c^2}{g}$$

We have to go out this distance from O to find an observer with acceleration g. The speed of light is $c = 3 \times 10^8$ meters per second, so c^2 is approximately 10^{17}. That gives R equal to about 10^{16} meters. Therefore we have to go out ten thousand billion kilometers to find an observer with approximately the acceleration we are familiar with on the surface of the Earth.

So let's go there!

And while there, if we don't move too much along the r direction,[6] the acceleration $g = c^2/R$ won't change much. It is similar to moving on a vertical axis near the surface of the Earth: the gravitational field doesn't change much.

We will now analyze what an observer in an elevator at position R is feeling when the elevator is accelerating uniformly (i.e., evolving in time along a hyperbola that is almost vertical). In figure 15, the spatial axis is horizontal, therefore the elevator is somehow on its side. But of course we should think of it as vertical – we are used to this presentation of the spatial axis in a Minkowski diagram. We will show that there is a fictitious gravity.

Since we are focusing on the vicinity of the point R, let's introduce a new spatial coordinate y measuring the distance from R.

In order to reach, in the vicinity of R, equations that look as much as possible like the Minkowski equations we are familiar with, we will also change the time variable.

---

[6]The r direction is nothing more than the X-axis in the Minkowski diagram of figure 14.

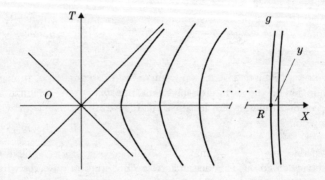

Figure 15: Hyperbolas at R and $R + y$. Both are almost vertical lines.

We define

$$y = r - R \tag{38}$$

All the observers with a small value of y have approximately the same acceleration g. Then we rewrite the metric of equation (37) using the new local coordinate y (note that $dr = dy$):

$$d\tau^2 = (R^2 + 2Ry + y^2)\, d\omega^2 - dy^2 \tag{39}$$

Let's present it as

$$d\tau^2 = \left(1 + \frac{2y}{R} + \frac{y^2}{R^2}\right) R^2\, d\omega^2 - dy^2 \tag{40}$$

We will simplify it by focusing on a limited region around R.

One more step concerning coordinates: as said, we also introduce a new time coordinate. We give $R\omega$ a new name: we call it t.

In order to simplify equation (40), let's observe that y/R is very small. Indeed, the quantity y is measured in meters, or perhaps kilometers, while R is huge.

The quantity y^2/R^2 is even much smaller. So we are going to keep y/R and neglect y^2/R^2. After all this dressing and adjusting, equation (40) becomes

$$d\tau^2 = \left(1 + \frac{2y}{R}\right) dt^2 - dy^2 \tag{41}$$

We end up with a metric that apart from the term $2y/R$, which is small, just looks like the good old Minkowski metric $dt^2 - dy^2$. It is space and time in a more or less ordinary way, but with a little corrective term. That little corrective term $2y/R$ is what accounts for gravitation in the accelerated reference frame. Keep in mind, though, that we are still in flat space, like when we traveled in the accelerated elevator of lecture 1. So far we have not dealt with any curvature.

With a change of coordinates, we could go back to the Minkowski metric defining a flat space. Therefore any gravitation that we find is, in a sense, the same fake gravitation that we found in lecture 1.

Let's step back to see where we are. We are studying physics in an accelerated coordinate system. It is the elevator being pulled toward the right (because on the usual diagram the spatial axis is horizontal). What do we expect to find? We expect to find that in that elevator there is an *effective gravitational field* (also called a fictitious gravitational field). It is this field that is associated with the term $2y/R$ in equation (41).

To get a better understanding of the connection, let's now study the motion of a particle in a metric given by equation (41). In units where c is equal to 1, we have $g = 1/R$. The metric can be rewritten

$$d\tau^2 = (1 + 2gy) \, dt^2 - dy^2 \qquad (42)$$

Have you ever seen the expression gy in studying gravitation in a uniform field? If we introduce the mass m of the particle, mgy is simply the potential energy. The term gy is called the *gravitational potential*. And the term $(1 + 2gy)$ is one plus twice the gravitational potential.

This is extremely general. In any kind of gravitational field, as long as it is more or less constant with time, and not doing anything too radically relativistic, the coefficient in front of dt^2 in the metric is always one plus twice the gravitational potential.

Why do we call gy the gravitational potential other than that it just looks like it? It is because if we work out the equation of motion of a particle in the metric given by equation (42), we will find that, as long as the particle is moving slowly, as long as we can make a good Newtonian approximation, as long as things are not too relativistic, the equation of motion that we will find is the same as that of a particle falling along the y-axis in a uniform gravitational field, as we have calculated it in classical mechanics; see volume 1 of TTM. What we mean by the y-axis of course is still the unique spatial axis, but near point R.

The whole point of the preceding section on uniform acceleration was to explain that a uniformly accelerated reference frame is something more subtle than just accelerating a system of observers along the X-axis. We had to *define* what we meant by a uniformly accelerated frame. It led to a construction where points at different distances from O, measured for instance at time 0, each have a fixed proper acceleration, but different from point to point. On the other hand, we can check with a Lorentz transformation, in the frame of one of the moving points at velocity v, that the distances measured simultaneously by P between the points don't change. It also led to hyperbolic coordinates and to figure 13.

Aside from the somewhat fancy looking mathematics, it is really fairly ordinary physics. There is an accelerated elevator at M, shown in figure 16. There is another accelerated elevator at N. There is another one at P. There is one at R, etc. The uniformly accelerated frame is just a collection of elevators at different positions, each being accelerated, but with a different acceleration. They have to be accelerated, according to equation (34), with acceleration c^2/R.

We are interested, in figure 16, in the accelerated elevator at hyperbolic radius R. Of course, let's stress once more that the elevator should be imagined on its side.

The bottom of the elevator follows the trajectory with hyperbolic radius R. It is the thick trajectory shown in figure 16.

The metric of space-time in the coordinate system (t, y), which we built to locate things inside the uniformly accelerated elevator, is given by equation (42) reproduced here:

$$d\tau^2 = (1 + 2gy)\ dt^2 - dy^2$$

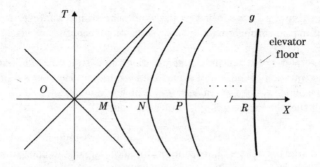

Figure 16: On the floor of the uniformly accelerated elevator at radius R, an observer experiences a fictitious gravity g. Everything is oriented horizontally on the diagram, but think of the X-axis as vertical.

We had to do some work to arrive at the last equation. Now it is going to be a given. It is the metric tensor, in the coordinate system $(t, \ y)$, at points in the vicinity of a point moving with acceleration g. In order to show the equivalence of a uniform acceleration and a gravitational field, we chose for the acceleration g the same value as the gravitational field of the Earth.

Motion of a Particle

What is the rule to figure out how a particle moves? The rule is that free particles move on *geodesics* – not geodesics of space but geodesics of space-time.

That is an important point that we proved with equation (19), and that is at the heart of general relativity. So let's stress it again:

In space-time, free particles always move along geodesics. It is true whether the space-time is flat or curved by masses.

When the space-time is curved by masses, the geodesics are no longer simple straight lines (when we use Minkowski coordinates). There are geometric distortions similar to those we represented in figure 5.

In other words, we take the metric of space-time, whatever it is, and we go through exactly these same operations.

The metric of space-time that we are given is equation (42). And the equation of motion of a particle is equation (19). Let's reproduce it with a change of sign and a new number:

$$-\frac{d^2 X^n}{dS^2} = \Gamma^n_{mr}\, t^r\, t^m \tag{43}$$

It is the equation of motion that says: move along a geodesic. Go straight ahead. But not straight ahead in space, *straight ahead in space-time*. Otherwise the equations are the same. We shall write it slightly differently, using $d\tau$ instead of dS. Remember that, when we work with $c = 1$, $d\tau^2$ is just the opposite of dS^2. So the left-hand side of (43) becomes

$$\frac{d^2 X^n}{d\tau^2} \tag{44}$$

This is called the *proper acceleration*. As long as the elevator is moving slowly, in other words if it hasn't been accelerating long enough to get near the speed of light, then the proper time and the ordinary time are essentially the same. And expression (44) is just the ordinary acceleration.

We choose X to be y. We want the y component of acceleration. Then expression (44) is simply

$$\frac{d^2 y}{d\tau^2}$$

Now we turn to the right-hand side of equation (43). The n-th component of X stands for y. What is t^r? It is dX^r/dS, because we are on a geodesic. And dS is $id\tau$. So equation (43) becomes

$$\frac{d^2 y}{d\tau^2} = -\Gamma^y_{mr}\, \frac{dX^r}{d\tau}\, \frac{dX^m}{d\tau} \tag{45}$$

Since m and r each run over four coordinates,[7] the right-hand side has a whole collection of terms – ten of them to be precise because

---

[7] To be consistent with standard notations in relativity, it would be better to use μ and ν. We leave it to the reader to make the change in these dummy variables to the usual ones.

the gammas are symmetric in m and r. Fortunately most of them are extremely small as long as the elevator is moving slowly, and as long as the movement of the object we are interested in, namely the particle with coordinate y, is slow. Under these conditions, only one of the combinations $\Gamma^y_{mr} \frac{dX^r}{d\tau} \frac{dX^m}{d\tau}$ is significant.

What is the value of $dt/d\tau$ for slow motion? It is essentially 1, because time and proper time in that case are almost the same.

On the right-hand side of equation (45), the differential elements are the components of the 4-velocity of the particle. We just saw that $\frac{dX^0}{d\tau}$ is essentially 1.

What are the derivatives of the space components with respect to τ? They are proportional to their actual ordinary spatial velocity. We are assuming that the spatial velocity is small compared to the speed of light, so the only important contribution on the right-hand side of (45) comes when r and m are time indices. Let's use t instead of 0 for the time index. Equation (45) reduces to

$$\frac{d^2 y}{d\tau^2} = -\Gamma^y_{tt} \tag{46}$$

The right-hand side must be the gravitational force. It must be the derivative of the gravitational potential energy.

Let's go back to the expression of the Christoffel symbols in terms of the metric as we saw in equation (5), which we reproduce here in a slightly different form:

$$\Gamma^p_{rs} = \frac{1}{2} \, g^{pn} \left[\frac{\partial g_{nr}}{\partial X^s} + \frac{\partial g_{ns}}{\partial X^r} - \frac{\partial g_{rs}}{\partial X^n} \right] \tag{47}$$

We need the symbol with two time covariant indices and one space contravariant index. The space index is y. Among the terms g^{yn}, the only one that is not negligible is g^{yy} and it is 1. Since X^t is just what we denote t and X^y what we denote y, we get

$$\Gamma^y_{tt} = \frac{1}{2} \left(\frac{\partial g_{yt}}{\partial t} + \frac{\partial g_{yt}}{\partial t} - \frac{\partial g_{tt}}{\partial y} \right)$$

The terms $\frac{\partial g_{yt}}{\partial t}$ and $\frac{\partial g_{yt}}{\partial t}$, which are equal, are both zero. So finally

$$\Gamma_{tt}^{y} = -\frac{1}{2} \frac{\partial g_{tt}}{\partial y}$$

Equation (46) can be rewritten

$$\frac{d^2 y}{d\tau^2} = \frac{1}{2} \frac{\partial g_{tt}}{\partial y} \qquad (48)$$

An equation like (48), where the second derivative of a spatial variable y with respect to time is proportional to the first derivative of some quantity with respect to y, reminds us of an equation of motion with potential energy. Somehow one half of g_{tt} must be the opposite of a potential energy. But we saw that indeed it is minus a potential energy with $m = 1$, also called gravitational potential.

In equation (42), reproduced here

$$d\tau^2 = (1 + 2gy)dt^2 - dy^2$$

g_{tt} is the coefficient $-(1 + 2gy)$ in the metric defining dS^2, which is the same as $d\tau^2$ with a minus sign. Therefore one half the derivative of g_{tt} with respect to y is $-g$. Equation (48) finally becomes

$$\frac{d^2 y}{d\tau^2} = -g \qquad (49)$$

That is the equation of motion of a particle in a uniform gravitational field. We went through a rather complicated derivation to reach it, but in so doing we learned the following points:

1. Space-time has a metric. In arbitrary coordinates, the metric can have a fairly complicated structure. In uniformly accelerated coordinates, however, it is almost the Minkowski metric, but with the extra term $2gy$ in equation (42).

2. The equation of motion along a geodesic in space-time – at least as long as things are going slowly, that is, as long as a Newtonian approximation is valid[8] – is just Newton's equation in a uniform gravitational field.

---

[8]This means setting the terms where c appears in the denominator to 0.

Uniform gravitational field, constant acceleration, equality with $-g$, etc., it is what we expected. But to analyze the physics properly, using the metric, the Christoffel symbols, the geodesics, and so forth, we followed a mathematically fairly heavy procedure.

Einstein guessed it. The hypothesis that a particle moves along a geodesic in space-time was his starting point, and he went in the opposite direction. He knew about uniformly accelerated coordinate systems, but he didn't know about Christoffel symbols. Somewhere along our own derivations is where he started. And for a uniform acceleration – with Newtonian approximation – the metric is simply given by equation (42).

We have come around full circle from the first lecture where we studied accelerated elevators in flat Newtonian space giving rise to some sort of gravitation. We have shown that in the Minkowski–Einstein space-time a uniformly accelerated reference frame similarly does give rise to an effective gravitational field.

But so far we haven't gotten to real gravitational fields. The gravitational field we observed is not a real gravitational field because our entire analysis took place in a flat space-time.

If we took the metric of equation (42) and calculated the curvature tensor, it would be exactly zero, indicating that there do exist coordinates where the metric has the simple form $dT^2 - dX^2$.

Thus the gravitation that we are experiencing is really exactly the gravitation due to an accelerated frame of reference, not to any real gravitating matter.

We can guess what the effect of real gravitating matter would be. Instead of $2gy$ in equation (42), what is the gravitational potential due to a gravitating object? It is $-G/y$.[9]

---

[9]By convention, for a uniform gravitational field, the gravitational potential is taken to be zero at ground level and increases to $+\infty$ when the height increases, while for the radial field created by an object, the gravitational potential is taken to be zero infinitely far away and goes to $-\infty$ when the radius goes to zero. That is why y is now in the denominator and there is a minus sign.

We can expect, when we study the metric of a real gravitational field, that we will have something looking like this:

$$d\tau^2 = \left(1 - \frac{2GM}{y}\right) dt^2 - dy^2 \qquad (50)$$

where G is Newton's constant and M is the mass of the gravitating object, such as the one depicted in figure 17.

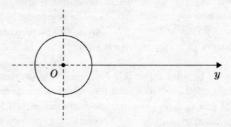

Figure 17: Gravitating object of mass M and gravitational potential $-G/y$.

That is almost the Schwarzschild metric, but not quite. We will work out what is the Schwarzschild metric of a gravitating object.

Equation (50) will lead to a weird phenomenon. When y is large, the term $2GM/y$ is small. That is good because $(1 - 2GM/y)$ is positive. But something crazy happens at the point where y is equal to $2GM$. The coefficient in front of dt^2 becomes zero. That point y where the coefficient changes sign is called the *horizon* of the black hole.[10]

Real gravitational fields and the Schwarzschild metric will be the subject of the next lecture. We are not going to derive the metric entirely from what we already know. To derive it, we need field equations. We haven't discussed them yet, and we won't do so until lecture 9.

So far we have only discussed geometry, flatness, curvature, geodesics, etc. When we finally arrived at the space-time of relativity

---

[10]For the Earth, supposing all its mass was almost point-like, the horizon would be 9 millimeters.

and its peculiar geometry, we ended up with a little demonstration of how, in a uniformly accelerated reference frame in space-time, movement along a geodesic gives rise to Newton's equation.

In lecture 5, we will finally be in a space-time where, as opposed to what we studied here, where we found only effective[11] curvature in a basically flat space-time, there will be gravitating masses creating real curvature of space-time.

We will begin by studying a black hole, and the metric of space-time created by such a thing, because a black hole is the simplest kind of massive object in general relativity – equivalent to a point mass in Newtonian physics.

---

[11]Where – we remember – *effective* is opposed to *real*.

Lecture 5:

Metric for a Gravitational Field

Andy: *Ah, we now come to this mysterious name: Schwarzschild. It has always puzzled me. Does it mean he is the son of a black hole?*

Lenny: *No. But Karl Schwarzschild did some fundamental work with Einstein equations that led to the hypothesis that black holes must exist.*

Andy: *But what are they?*

Lenny: *Be patient! They are coming. However, we have some work to do first.*

Andy: *I can see them looming on the horizon.*

Time-like, space-like, and light-like intervals and light cones
Geodesics and Euler–Lagrange equations
Schwarzschild metric
Black holes
Event horizon of a black hole
Motion of a light ray

Time-like, Space-like, and Light-like Intervals and Light Cones

Let's begin with time-like, space-like, and light-like intervals. For that we go back to special relativity to spell out what that means. We have discussed the metric many times. We call it proper time. The square of the proper time is defined as

$$d\tau^2 = dt^2 - dx^2 - dy^2 - dz^2 \tag{1}$$

Remember that equation (1) is the expression of the proper time when we work with units in which the speed of light is $c = 1$. Its full expression making the speed of light explicit is

$$d\tau^2 = dt^2 - \frac{1}{c^2}\left(dx^2 + dy^2 + dz^2\right) \qquad (2)$$

One can also use, for the four variables of space-time, t, x, y, z, the more relativistic notations, X^0, X^1, X^2, X^3. Equation (1) then becomes

$$d\tau^2 = \left(dX^0\right)^2 - \left(dX^1\right)^2 - \left(dX^2\right)^2 - \left(dX^3\right)^2 \qquad (1')$$

And equation (2) becomes

$$d\tau^2 = \left(dX^0\right)^2 - \frac{1}{c^2}\left[\left(dX^1\right)^2 + \left(dX^2\right)^2 + \left(dX^3\right)^2\right] \qquad (2')$$

Let's also recall that when we use the notation X^m with a Latin index, we mean the 3-vector of spatial coordinates $\left(X^1,\ X^2,\ X^3\right)$, while when we use the notation X^μ with a Greek index, we mean the full 4-vector $\left(X^0,\ X^1,\ X^2,\ X^3\right)$, where the first coordinate, with index 0, is time.

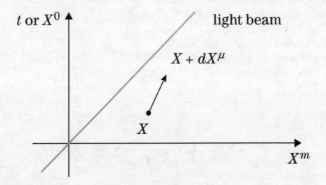

Figure 1: Flat space-time in the usual Minkowski diagram, and a small displacement represented by the 4-vector dX^μ.

The rationale to sometimes introduce the speed of light explicitly in equation (2), or its variant with the relativistic notation (2'), is to keep track of what is small and what is big under certain circumstances. For example, if we want to go to the non-relativistic limit, that is, the limit where everything is moving slowly, it is good to put back c because it reminds us that it is much bigger than any other velocities in the problem. And it makes it easy to see which terms can be neglected and which cannot.

In what follows, unless it is necessary to show explicitly c, we will set it equal to 1. Notice that in the standard Minkowski diagram of special relativity, because we take $c = 1$, light rays have a slope at 45°.[1] When we represent two spatial dimensions beside the time dimension, light cones have a generatrix tilted at 45°.

Let's look at the sign of $d\tau^2$. Of course, when we look at real numbers, their square is always positive. But $d\tau^2$ is not defined as the square of a real number, it is defined by equation (1) or (2). It can be positive, null, or negative, depending on whether $dx^2 + dy^2 + dz^2$ is smaller than dt^2, equal to it, or bigger than it.

If $d\tau^2 > 0$, then the little element dX^μ in figure 1 on the preceding page is said to be *time-like*. It contains more time than it contains space, so to speak. Its vertical component is bigger than its horizontal component. Its slope is greater than 45°.

dX^μ

Figure 2: Time-like interval.

The time-like nature of dX^μ, in this case, can also be described in terms of a light cone, shown in figure 2. If we represent two spatial coordinates x and y, in addition to the time coordinate t, and a light cone whose center is at X, then $d\tau^2 > 0$ means that the little 4-vector dX^μ lies in the interior of the cone. It could

[1]One also meets an alternative diagram where instead of t the vertical axis charts ct, and light rays are still at 45° irrespective of the units.

also lie in the backward direction in the same picture, pointing to the past. Either way, dX^μ is called a time-like interval.

Space-like is exactly the opposite of time-like. It corresponds to $d\tau^2 < 0$, or equivalently $dx^2 + dy^2 + dz^2$ greater than dt^2. In that case, we usually define another quantity dS called the proper distance. Let's temporarily reintroduce explicitly c to recall what it is. By definition the square of the proper distance is

$$dS^2 = dx^2 + dy^2 + dz^2 - c^2 dt^2 \qquad (3)$$

When we are in units where $c = 1$, dS^2 has the same form as $d\tau^2$, shown in equation (1), except for a change of signs. Otherwise, in general, we have $dS^2 = -c^2 d\tau^2$.

Space-like vectors are those for which $dS^2 > 0$. If we represent as before the cone at X, a space-like little interval dX^μ is shown in figure 3.

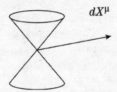

Figure 3: Space-like interval.

Finally there are *light-like* vectors. They are those for which $d\tau^2 = 0$, and therefore equivalently $dS^2 = 0$. In the standard diagram, their slope is at 45°. They are trajectories of light rays, and they lie on the surface of the cone of figure 3.

Those are the three kinds of 4-vectors in Minkowski space.

Just for a moment, consider what it would mean if there were two positive signs and two negative signs instead of one positive sign and three negative signs in the definition of the metric. This would correspond to two time dimensions. It doesn't mean anything in physics. There are never two time dimensions. There is always one time and three space dimensions. Can you imagine a world with two times? Personally I cannot imagine what it would mean to have two different time dimensions. So we will simply take the view that it is not an option: there is always one time-like dimension in the metric of equation (1), or its variant forms, and three space-like.

That doesn't mean that there is a unique *direction* that is time-like. There are many time-like directions pointing within the light cone of figure 2.

The invariant property, corresponding to the fact that at any point there is one time and three space variables, concerns the metric tensor. We are familiar with the expression of the metric tensor as follows:

$$d\tau^2 = -g_{\mu\nu} \, dX^\mu \, dX^\nu \tag{4}$$

The minus sign in front of $g_{\mu\nu}$ is a convention. When we choose to use dS^2, the square of the infinitesimal *proper distance*, it is given by the same expression but without the minus sign.

Let's stress the following important point about the *proper time* $d\tau$, whose square is defined by equation (4): it is the time recorded by a clock accompanying the particle along its trajectory – its wristwatch if you will. In other words, it has a physical practical meaning, which is often useful to remember. Of course, for particles going slowly – and by "slowly" we mean up to thousands of miles per second – the proper time is essentially the same as the standard time t of the stationary observer in the stationary frame of figure 1. This is easily derived from equation (2), because c is very big compared to ordinary velocity, or to the spatial components of the 4-velocity. (Go to volume 3 of TTM on special relativity, if you need to brush up on these ideas.)

Similarly, the *proper distance* dS, along the trajectory of a particle, is distance measured by a meter stick carried along by the particle.

In summary, proper distance, $\sqrt{dS^2}$, really is a distance. And proper time, $\sqrt{d\tau^2}$, really is a time. Let's keep that in mind.

Equation (3) is the definition of the metric with the coordinates (t, x, y, z). It can always be written in terms of a matrix. In this case (and we are back to $c = 1$), it is the matrix η shown here

$$
\eta_{\mu\nu} = \begin{pmatrix} -1 & 0 & 0 & 0 \\ 0 & 1 & 0 & 0 \\ 0 & 0 & 1 & 0 \\ 0 & 0 & 0 & 1 \end{pmatrix} \tag{5}
$$

Remember that it is the analog in Minkowski space of the Kronecker-delta, which is simply the unit matrix, in Euclidean space.

The matrix η has obviously three positive eigenvalues and one negative eigenvalue. That is the invariant story: *there is always only one negative eigenvalue in the metric*. In special relativity as well as in general relativity, that will still be the case whatever the metric is and whatever the coordinate system is. Since the metric, in general, depends on the point in space-time where we look at it, this invariance statement will be true at any point.

A metric that would have two negative eigenvalues or three negative eigenvalues, would have more than one time *dimension* (not to be confused with a time-like direction, of which there are many – they are those inside the Minkowski cone[2]). We just don't even think about several time *dimensions*. Several time axes is something that physics does not seem to have a use for.

The concepts of time-like, space-like, and light-like displacements are not restricted to special relativity. They apply generally whatever the metric, and whatever the point we consider. In the preceding discussion, we were in the flat space of special relativity, but the concepts apply in general relativity where the space is intrinsically not flat.

Now we shall consider a metric more general than $\eta_{\mu\nu}$. We denote it $g_{\mu\nu}(X)$. At every X (i.e., every event in space-time), there is a matrix. Furthermore that matrix must have one negative and three positive eigenvalues. In other words, wherever you stand, you should experience a world with one time dimension and three space dimensions, or more exactly a metric with one negative eigenvalue and three positive.

---

[2] Another name for the light cone.

That means that every point in space has a light cone associated with it; see figure 4. These light cones can be tilted and change shape depending on the curvy aspect of the coordinates at each point.

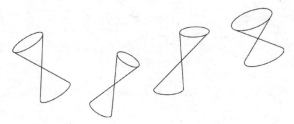

Figure 4: Light cone at each point.

But at each point the metric has three positive and only one negative eigenvalue. And at each point there is the notion of time-like displacement, space-like displacement, and light-like displacement.

The property of having a certain number of eigenvalues positive and a certain number negative is called the *signature* of the metric. Recall that we already presented the concept of signature in lecture 4 (in the paragraph following equation (23) of that lecture). What is the signature of the metric of ordinary flat space? I'm not talking about ordinary space-time, I just mean the page you are reading. It is $+\,+$. And the signature of the Minkowski metric in special relativity with three spatial coordinates is $-\,+\,+\,+$.

When somebody gives us a metric, or wherever we get it from – we might get it as a present in the mail, or we might have calculated it from some equations of motion, or some field equations – we should make sure that that metric has the signature $-\,+\,+\,+$. If it doesn't, it means something's wrong. Moreover, not only should we have that signature at some point, but we should have it at every point in space-time.

Notice that the shape of the light cone, in particular its angle of openness, is a pure coordinate issue; see figures 2, 3, and 4. In particular, if in the standard Minkowski metric and representation of figure 2, we chose units such that the speed of light is not 1, but for instance the huge number 3×10^8, the cone would be extremely flat, and the picture not very useful. We already mentioned that in volume 3 on classical field theory and special relativity.

So much on the signature of the metric.

In this lecture, our ultimate objective is to study the metric around a massive body, and more specifically around a special kind of massive body: a black hole. But first, let's revisit geodesics in space-time, deriving them in a different way than what we did before.

Geodesics and Euler–Lagrange Equations

We learned in lecture 4 the definition of a geodesic in space-time.[3] We used the corresponding equation – equation (19) of lecture 4 – in the example of a free particle moving in a uniformly accelerated reference frame. A geodesic is a curve whose tangent vector stays parallel to itself all along the curve. Said in another, more informal way, it is a trajectory where we always go straight.

In this lecture, we are going to use a different definition, which in many ways is more useful. But let us first recall the original definition:[4]

$$\frac{d^2 X^\mu}{d\tau^2} = -\Gamma^\mu_{\sigma\rho} \frac{dX^\sigma}{d\tau} \frac{dX^\rho}{d\tau} \tag{6}$$

The left-hand side is the derivative of the tangent vector along some curve, which all along the curve should be equal to the double sum involving Christoffel symbols of the right-hand side. That is the standard definition of a geodesic.

---

[3] A geodesic in ordinary curved space, i.e., Riemannian space, is a rather intuitive concept: it is a curve that minimizes ordinary distance. On Earth, for instance, it is a portion of great circle. In space-time it is less intuitive. We already gave a technical definition related to "straightness," and mentioned others in lecture 4. We will now study it as the result of a minimization problem.

[4] It was equation (19) of chapter 4. Here we write it with τ instead of S, and with the more explicit expressions for the tangent vector components on the right-hand side.

Remember that in lecture 4, we also mentioned another: it is the analog of the definition of a geodesic in ordinary space, that is, the curve of shortest distance between two points. Or better yet, it is the curve between two points whose length is stationary.

Remember too that on a geodesic the covariant derivative of the tangent vector is everywhere zero.

Another way to arrive at equation (6) is to "extremelize" – by that I mean make minimum – the length of the curve between two points.

Let's start with ordinary space to refresh our memory on geodesics. So we are on the page of this book, or a curved version of the page with hills and valleys, for instance after it stayed in the rain and then dried. We take two points in that space and any curve between them, as shown in figure 5. We calculate the distance along the curve.

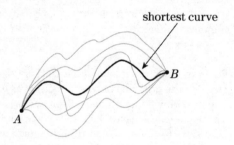

Figure 5: Determination of a geodesic between A and B. When the space is flat, it is the straight segment joining them.

Then we search for the curve that minimizes its length, denoted *shortest curve* on the figure.

How to calculate it? Let's spell out *the logic*. We start, as said, with any curve C between A and B. There are plenty of them shown in grey in figure 5. On that selected curve C, whichever it is, for each little segment along the curve, we have

$$dS^2 = g_{mn}(X)\, dX^m\, dX^n \tag{7}$$

$$dS = \sqrt{g_{mn}(X)\ dX^m\ dX^n} \tag{8}$$

This is just Pythagoras theorem applied to a little segment on curve C. Then we add them all up. This gives the distance along the curve C we have selected:

$$S = \int_{\text{along curve } C} \sqrt{g_{mn}(X)\ dX^m\ dX^n} \tag{9}$$

Finally we look for the curve C that makes S minimum or extremum. That's the logic. By now we are familiar with the mathematics that implements this logic. We learned it in volume 1 of TTM on classical mechanics. It is a problem in calculus of variation, analogous to minimizing the action of a particle along a trajectory.

In other words, we can think of equation (9) as expressing the action of a particle moving from A to B along curve C. Then the rule for calculating the geodesic is to "extremelize" that quantity, or more accurately to make it stationary. The equation that tells us how to minimize a quantity like S in equation (9) is called the Euler–Lagrange equation.

When we go from the principle of least action to the Euler–Lagrange equation, the principle of least action turns into a differential equation involving a Lagrangian. Typically, when rewritten as explicitly as possible, the Euler–Lagrange equation becomes an equation of the type $F = ma$. Going from minimizing the quantity in equation (9) to equation (6) is exactly the same operation. In fact equation (6) looks like equating an acceleration to something. That thing is a kind of force.

Now let's come back to relativity and to our actual problem of geodesic, where we are not concerned with ordinary distance but with proper distance, or equivalently proper time.

If we want to express the quantity to be minimized, which involves proper time, we deal with almost exactly the same expression as in equation (9) except for a minus sign in front of the metric. From equation (4) we find that the proper time between point 1 and point 2 in space-time is given by

$$\tau = \int_1^2 \sqrt{-g_{\mu\nu}(X)\ dX^\mu\ dX^\nu} \qquad (10)$$

This is the expression that we will want to "extremelize".

Notice that a time-like geodesic *maximizes* proper time (it is one of the explanations of the twin paradox). The usual definition of action is proportional to *minus* the proper time:

$$\text{action} = -m \int d\tau$$

Minimizing action means maximizing expression (10).

Let's suppose that the expression defined by equation (10) really corresponds to the motion of a particle that starts at point 1 and ends at point 2 in space-time, as in figure 6.

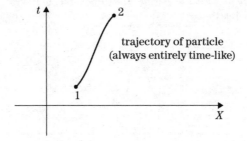

Figure 6: Trajectory of a particle: geodesic in space-time.

The action we are interested in depends on one more quantity. It depends on the mass m of the particle. The actual action then is

$$A = -m \int_1^2 \sqrt{-g_{\mu\nu}(X)\ dX^\mu\ dX^\nu} \qquad (11)$$

This is a definition of the mass. We will find out that putting a coefficient called mass here is important for thinking about energy and so forth; and the minus sign is strictly a convention in the definition of mass. We want to make this action A stationary.

What do we do with the right-hand side of equation (11)? A priori it is a completely unrecognizable object to work on with our
mathematical toolbox. That is, it is just a sum of infinitely many
infinitely small elements.[5] But where is the differential element?
What is the variable to integrate over?

Usually for us, an integral we know how to calculate, or at least
manipulate, has the form

$$\int F(\text{ some variable }) \, d \text{ some variable}$$

where the variable may be some spatial quantity or may be time,
or some other clearly identified physical quantity. Normally we
don't see integrals where beneath the integral there is a square
root, and inside the square root there is a product of differentials
like $dX^\mu \, dX^\nu$.

Remember that we already met the same kind of integral in volume 3
on special relativity.

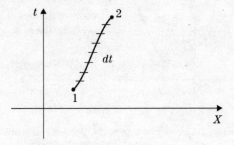

Figure 7: Breaking the trajectory into little time segments dt.

To start with, let's break up the trajectory of the particle into
little time segments dt, as in figure 7. Equation (11) becomes

$$A = -m \int_1^2 \sqrt{-g_{\mu\nu}(X) \, \frac{dX^\mu}{dt} \, \frac{dX^\nu}{dt} \, dt^2} \qquad (12)$$

---

[5]Mathematically speaking, it is more precisely *defined* as the limit, when
the number N of elements goes to infinity and the sizes of the elements go to
zero, of finite sums of N elements of the form $f(u_n)\Delta u_n$. Such a definition
is also due to Riemann.

Some of the differentials dX^μ are in fact dt, because t is one of the four coordinates in $X^\mu = (t,\ x,\ y,\ z)$. Remember that we set $c = 1$; otherwise X^μ would be $(ct,\ x,\ y,\ z)$. That would produce a coefficient c^2 in factor next to the mass m. We will reinsert c^2 in the next section when we study the Schwarzschild metric.

What happens when we have dt/dt? It is just 1. And what happens when we have dx/dt or the analog with y or z? It is just the ordinary velocity. We can also pull dt^2 out of the square root and obtain a standard differential element dt in the integral:

$$A = -m \int_1^2 \sqrt{-g_{\mu\nu}(X)\, \frac{dX^\mu}{dt}\, \frac{dX^\nu}{dt}}\ dt \qquad (13)$$

At each time t, the quantity $\sqrt{-g_{\mu\nu}(X)\, \frac{dX^\mu}{dt}\, \frac{dX^\nu}{dt}}$, which is integrated over time, has a definite value along the trajectory. It is a certain function of the velocity and the position X. Thus we have transformed our expression (11) for the action into a conventional integral over time along the trajectory in figure 7.

The integrand in equation (13) (where we put m back beneath the integral sign) is the Lagrangian. It is the quantity that, in the calculation of a geodesic, plays exactly the same role as the Lagrangian when we apply the principle of least action to calculate the trajectory of a particle in classical non-relativistic physics. Action, by definition, is equal to the integral of the Lagrangian, which is itself a function of velocities and positions:

$$A = \int \mathcal{L}(\dot{X},\ X)\, dt \qquad (14)$$

In summary, "extremelizing" the action given by equation (13) brings us back to a problem that we already met in classical mechanics, in volume 1. How do we find an equation of motion from an action? In order to do that, we solve the Euler–Lagrange equation (or equations) that the Lagrangian must satisfy.

In our present problem, the Lagrangian is

$$\mathcal{L} = -m \sqrt{-g_{\mu\nu}(X)\, \frac{dX^\mu}{dt}\, \frac{dX^\nu}{dt}} \qquad (15)$$

Incidentally, is the quantity inside the square root positive?[6] Can we take its square root, even though there is a minus sign in front of $g_{\mu\nu}$? Answer: yes, it is positive. The quantity $-g_{\mu\nu}(X)\, dX^\mu\, dX^\nu$ is the square of the proper time over a small element dX along the trajectory; see equation (1). It is always positive for a time-like trajectory. And, as we already said in lecture 4 (as well as in many places in volume 3), particles always move on time-like trajectories. It is equivalent to saying that they never exceed the speed of light.

This is a point worth stressing:

Particles do not move faster than the speed of light.

Therefore, on the standard Minkowski diagram, the trajectory of a particle never has a tangent with a slope lower than 45°. The simplest example of course is a particle that doesn't move in the referential of the Minkowski diagram: its trajectory in space-time is vertical.

We also saw that fact with the hyperbolas on which a collection of observers were moving simultaneously when we studied the concept of uniform acceleration in lecture 4. That is why in figures 6 and 7 we were careful to draw curves with tangents always higher than 45°.

To finish on this remark: a space-like trajectory would be one where the point moves faster than the speed of light. But this is impossible. Thus on a space-like interval (for instance the simplest one: a horizontal segment), we necessarily see many different particles (this is obvious also because the time is the same).

Let's recall what are the Euler–Lagrange equations that the Lagrangian must satisfy. First of all we are going to partially differentiate $\mathcal{L}$ with respect to each of the variables $\dot{X}^\mu$. But the first of these variables, which is the derivative of time with respect to time, is just 1. There is no corresponding equation. We are only

---

[6]Remember that when we deal with actions and Lagrangians, we are dealing with real quantities, for which the notion of maximizing or minimizing has a meaning. This would not be the case if we were dealing with complex numbers.

concerned with the three partial derivatives of $\mathcal{L}$ with respect to the components of ordinary velocity. On the left-hand side, for each of these partial derivatives, we take the derivative with respect to time and equate it to the partial derivative of $\mathcal{L}$ with respect to the corresponding component of position. Therefore the Euler–Lagrange equations are the three equations

$$\frac{d}{dt}\frac{\partial \mathcal{L}}{\partial \dot{X}^m} = \frac{\partial \mathcal{L}}{\partial X^m} \tag{16}$$

where m runs from 1 to 3. We learned them in classical mechanics.

The point is that if you know the metric $g_{\mu\nu}(X)$, from equation (16) you can work out an equation of motion for the particle. The particle's motion will be a geodesic in the sense of the trajectory with smallest action when integrated along the curve.

How does this relate to the definition of a geodesic given by equation (6)? Answer: if you work out exactly equation (16) with the given metric $g_{\mu\nu}(X)$, you will discover that you arrive precisely at equation (6).

Exercise 1: Given the metric $g_{\mu\nu}(X)$, show that the Euler–Lagrange equation (16) (we drop the "s"), to minimize the action along a trajectory in space-time,

$$\frac{d}{dt}\frac{\partial \mathcal{L}}{\partial \dot{X}^m} = \frac{\partial \mathcal{L}}{\partial X^m}$$

where the Lagrangian $\mathcal{L}$ is

$$\mathcal{L} = -m\sqrt{-g_{\mu\nu}(X)\,\frac{dX^\mu}{dt}\,\frac{dX^\nu}{dt}}$$

is equivalent to the definition of a geodesic given by equation (6), which says that the tangent vector to the trajectory in space-time stays constant:

$$\frac{d^2 X^\mu}{d\tau^2} = -\Gamma^\mu_{\sigma\rho}\,\frac{dX^\sigma}{d\tau}\,\frac{dX^\rho}{d\tau}$$

In general, working with the action defined by equation (13), and with the Euler–Lagrange equations (16), which come from minimizing this action, is much easier than working with a geodesic defined by equation (6).

We are going to work out some equations of motion for particles in a particular metric. The metric will be everybody's favorite metric: the Schwarzschild metric.

Schwarzschild Metric

We come to the questions of the metric of space-time and the motion of a particle in a *real gravitational field*, the gravitational field of the Sun, the Earth, or another massive spherically symmetric object.

X
•

Figure 8: Massive object and its gravitational field, and a point X in space at some distance from the object.

Assume that we are outside the mass of the object, at some distance from it, like in figure 8. We are interested in the metric of space at point X. First we are going to write a formula for the metric, and then we shall verify that the formula really does make sense. We mean that, using the equations of motion (16), it would give rise to something that looks very familiar, namely Newton's equations for a particle moving in a gravitational field – at least when we are far away from the gravitating object, where the gravitation is fairly weak.

Here is the metric we are going to use. First of all, if we were in a flat space, making c explicit, it would have the form

$$d\tau^2 = dt^2 - \frac{1}{c^2} \, dX^2 \qquad (17)$$

where henceforth dX^2 stands for $dx^2 + dy^2 + dz^2$. We expect indeed, as we go far away from the gravitating object, that the space-time there will look flat.

But we know that the gravitating object does something to space-time – warps it. So equation (17) should not be quite right for the metric of the space-time created by the object. It should be right only in the limit when we are far away. Therefore we add a coefficient in front of dt^2, and we leave the part with dX^2 as is:

$$d\tau^2 = \left(1 + \frac{2U(X)}{c^2}\right) dt^2 - \frac{1}{c^2}\, dX^2 \tag{18}$$

We guessed in lecture 4 (equation (50) of that lecture) that $U(X)$ is the gravitational potential due to the massive object in figure 8.

Let's check that, as long as our particle is moving slowly, its geodesic equation, in the Lagrangian form (16), just becomes Newton's equation for a particle moving in a gravitational potential $U(X)$.

The general form of the Lagrangian is given by equation (15), where inside the square root there is just $d\tau^2$. Now that we know the metric, from equation (18), we can write the Lagrangian or the action more explicitly:

$$A = -mc^2 \int \sqrt{\left(1 + \frac{2U(X)}{c^2}\right) dt^2 - \frac{1}{c^2}\, \frac{dX^2}{dt^2}\, dt^2} \tag{19}$$

Noting that $\frac{dX^2}{dt^2}$ is $\dot{X}^2$, after some rearranging, this becomes

$$A = -mc^2 \int \sqrt{\left(1 + \frac{2U(X)}{c^2}\right) - \frac{\dot{X}^2}{c^2}}\, dt \tag{20}$$

When we work with c explicitly present in the formulas, as we said, the expression for the action carries a c^2 next to the mass. We are familiar with that. It can be derived from a dimensional analysis, or we can remember that the action has units of energy multiplied by time.

We could work with the Lagrangian (putting back $-mc^2$ inside the integral) as it is, but what we are interested in is what happens when we move with a speed that is slow compared to c. Therefore the next step is to approximate the Lagrangian, and look at the Euler–Lagrange equations with the approximate Lagrangian.[7]

We are going to study these equations in the non-relativistic limit where the speed of light is huge. It will involve some approximations that we will spell out as we proceed.

We are interested in slow motion because we want to show that equation (20) really gives rise to Newton's equations in the non-relativistic limit. It is the limit where c is taken to be very large. This is the reason why in this section we chose to make c explicit.

Inside the square root, we can reorganize the terms as one plus a small quantity:

$$\sqrt{1 + \frac{1}{c^2}\left(2U - \dot{X}^2\right)} \tag{21}$$

Next we use the beginning of the binomial theorem:

$$\sqrt{1 + \epsilon} \approx 1 + \frac{\epsilon}{2}$$

Indeed, the quantity $\frac{1}{c^2}(2U - \dot{X}^2)$ is small because we are looking at a situation, as said, where c^2 is much bigger than $(2U - \dot{X}^2)$.

Thus expression (21) can be approximated by

$$1 + \frac{1}{2c^2}\left(2U - \dot{X}^2\right) \tag{22}$$

Going back to the expression for the action in equation (20), we get

$$A = \int \left(-mc^2 - mU + \frac{m}{2}\dot{X}^2\right) dt \tag{23}$$

---

[7]As usual, we rely on the mathematical fact that, in this case, it is okay to approximate the Lagrangian first and then to solve the Euler–Lagrange equations rather than solve properly the Euler–Lagrange equations first and then look at the limit case when $\dot{X}$ is small. In other words, here we can reverse the order of two big operations.

Inside the integral, we have the Lagrangian when the speed of the particle is small and we can make a non-relativistic approximation.

If we use this Lagrangian in the Euler–Lagrange equations, the constant term $-mc^2$ has no effect. The only thing we do with a Lagrangian is differentiate it. When we differentiate a constant, we get zero. So $-mc^2$ can be disregarded.

The other two terms in the Lagrangian are a conventional kinetic energy, $\frac{m}{2}\dot{X}^2$, minus a potential energy, $mU(X)$, which actually depends on X. For the potential energy we can choose the function we like. Incidentally, in a gravitational problem the potential energy of a particle is always proportional to its mass.

Finally, when we use this Lagrangian, the Euler–Lagrange equations will of course simply produce Newton's equation for a particle in a gravitational field $U(X)$, just as it did when we carried out exactly the same calculations in volume 1.

We arrive at

$$m\ddot{X} = -m\frac{\partial U}{\partial X} \tag{24}$$

On the right-hand side it is a force. On the left-hand side, it is an acceleration multiplied by the mass. The mass cancels.

The main point here is that the action given by equation (13), which is equivalent to the geodesic given by equation (6), is easily worked out using the Euler–Lagrange method. It is even easier to work out if we are in the non-relativistic limit, where we say that c is very large and $1/c^2$ is very small, and we can simply expand the square root in the expression of the action.

The important point that we learned carrying out these calculation is that, at least in some first approximation, we can write

$$-g_{00} = \left(1 + \frac{2U(X)}{c^2}\right) \tag{25}$$

In fact that turns out to be exact for g_{00}.

Of course, in equation (18), which we wrote for $d\tau^2$, there could be even smaller terms than $1/c^2$, terms in $1/c^4$ or $1/c^6$, etc. But they would be insignificant in the non-relativistic limit.

In short, we cannot say with complete confidence that $-g_{00}$ is one plus twice the potential energy of a particle divided by c^2. But we can say that it must be true to the first order in small quantities – small quantities meaning quantities with $1/c^2$.

Now for the gravitational potential energy created in space by a body of mass M, as in figure 8, we shall use

$$U(X) = -\frac{MG}{r} \tag{26}$$

where G is Newton's constant and r is the distance away from the center of the body.

We can write down our first guess at the metric of space-time surrounding a gravitational mass in figure 8. It is

$$d\tau^2 = \left(1 - \frac{2MG}{c^2 r}\right) dt^2 - \frac{1}{c^2}(dx^2 + dy^2 + dz^2) + \dots \tag{27}$$

where $r = \sqrt{dx^2 + dy^2 + dz^2}$. The three dots after the spatial terms stand for smaller things, one or more orders of magnitude smaller than $1/c^2$.

Readers familiar with cosmology[8] may recognize the Schwarzschild metric, or the metric of a black hole – but not quite.

Let's turn our attention to the term

$$dx^2 + dy^2 + dz^2 \tag{28}$$

It is the ordinary metric of 3D space. In equation (27) defining the metric of space-time created by a massive body, we have only fiddled with the time-time component of the metric so far. The space-space components of the metric and everything else have not yet been studied in detail. They will in a moment.

---

[8]Cosmology is the subject of volume 5 of TTM.

The space-space components in equation (27) are the metric of the ordinary flat space. Let's consider flat space in three-dimensional polar coordinates. These coordinates are characterized by a radius, namely the distance from the center of, say, the Sun, if we think of the body in figure 8 as the Sun, and a pair of angles. They can be a polar angle and an azimuthal angle, as shown in figure 9.

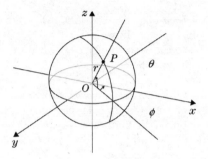

Figure 9: Spherical polar coordinates.

Coordinate r is the distance from the Sun O to P. The angle θ can be taken from the pole or from the equator. Let's measure it from the equator like on a terrestrial globe. It is the latitude. And the azimuthal angle ϕ is the longitude (or its opposite).

How is the ordinary 3D Euclidean metric $dx^2 + dy^2 + dz^2$ expressed in spherical polar coordinates? The formula, that the reader may have learned in trigonometry, is

$$dx^2 + dy^2 + dz^2 = dr^2 + r^2 \left(d\theta^2 + cos^2\theta \ d\phi^2\right) \qquad (29)$$

Despite its complicated look, it is simply the square of the length element on the surface of the sphere, $r^2 \left(d\theta^2 + cos^2\theta \ d\phi^2\right)$ that we have already met, to which is added dr^2 to complete Pythagoras theorem in three dimensions.

Why are we considering polar coordinates? Simply because polar coordinates are the most appropriate coordinates for studying a central force problem. So we are going to express the metric given by equation (27) in polar coordinates.

Let's introduce a convenient notation: we give $d\theta^2 + cos^2\theta \; d\phi^2$ a name, so as not to repeatedly write it out. We call it $d\Omega^2$. There is no other reason than to avoid writing it all the time in its full expression with θ and ϕ. Equation (27) can be rewritten

$$d\tau^2 = \left(1 - \frac{2MG}{c^2 r}\right) dt^2 - \frac{1}{c^2} \; dr^2 - \frac{1}{c^2} \; r^2 \, d\Omega^2 + \dots \qquad (30)$$

The variable r is the radial distance. And $d\Omega^2$ is the metric on the surface of the unit sphere. If we keep r fixed, the term $r^2 \; d\Omega^2$ is the contribution to the space-time metric when we move by some angles θ and ϕ on the sphere.

Now if we look at this metric, be it with expression (27) in Cartesian coordinates or with (30) in polar coordinates, there is something terribly wrong. It is fine when we are far away. But there is something deeply disturbing when we get close to the center.

What happens when r decreases? At some point, $2MG/c^2 r$ becomes bigger than one, and the coefficient

$$\left(1 - \frac{2MG}{c^2 r}\right)$$

becomes negative. In other words, the time-time part of the square of the proper time $d\tau$, the coefficient $-g_{00}$, becomes negative.

Since the other terms, which are the space-space parts in the definition of $d\tau^2$ given by equation (30) or (27), are also negative, we have a problem. The matrix $g_{\mu\nu}(X)$ expressing the metric tensor then only has positive terms on its diagonal, the terms off diagonal being all zero. Therefore it has four positive eigenvalues instead of one negative and three positive as it should.

Where does the problem arise? It arises where $\left(1 - 2MG/c^2 r\right)$ changes sign, that is, when

$$r \geq \frac{2MG}{c^2} \qquad (31)$$

The quantity $2MG/c^2$ is just some positive number, specific of the body we are looking at in figure 8. The metric no longer has three positive eigenvalues and one negative. Of course, this happens if all the mass of the body is somehow point-like.[9]

This leads us to the topic of black holes.

Black Holes

A black hole is a massive body whose mass M is *extremely* concentrated in a very small sphere. It is concentrated in a sphere of radius smaller than the value $2MG/c^2$, which we discovered to be the cause of a problem in the metric.

$2MG/c^2$ is a distance characteristic of the black hole. It is called the *Schwarzschild radius* of the black hole. Later we will see that it is also the radius of the *horizon* of the black hole.

When we approach a black hole, for instance radially, at some point, which is when we cross the Schwarzschild radius, the coefficient $-g_{00}$ changes sign and becomes negative, or equivalently g_{00} becomes positive, and the whole metric of equation (30) now has four positive eigenvalues.

The meaning of this is that we somehow entered into a region where there are four space directions and no time direction. That is bad! That is something we don't want. General relativity, or simply physics, doesn't allow that.

What really happens? When we study the field equations of general relativity in lecture 9, we will discover what really happens. For the time being, let me just say that at the point where the quantity $\left(1 - 2MG/c^2 r\right)$ changes sign, the rest of equation (30) will also change. We shall put an extra coefficient in front of the term dr^2/c^2 in equation (30) to the effect that when the time-time component of the tensor changes sign, the r-r component will also change sign.

---

[9]As the reader knows, if we enter inside Earth, we will only be submitted to the gravitational field created by the mass at a radius shorter than where we are, the mass farther away than us, within Earth, will play no role.

So what happens? The term $\left(1 - 2MG/c^2r\right) dt^2$ turns into a space direction, but simultaneously the term with dr^2 turns into a time direction. The signature of the metric is saved: one time-like dimension and three space-like dimensions.

What can we put, in equation (30), in front of the term dr^2/c^2 to make it flip sign when the time-time component flips sign? Well, we could also put $\left(1 - 2MG/c^2r\right)$ in front of dr^2/c^2. So if we crossed the Schwarzschild radius, the first term would change sign and the dr^2 term would too. But that turns out not to be quite the right thing.

Einstein's field equations give us a different answer. Instead of the same coefficient as in front of dt^2, use its inverse. From now on, disregarding the smaller terms in $1/c^4$ or more, we will use the following metric. It is the Schwarzschild metric in all its glory:

$$dr^2 = \left(1 - \frac{2MG}{c^2r}\right) dt^2 - \left(\frac{1}{1 - \frac{2MG}{c^2r}}\right) \frac{1}{c^2} \, dr^2 - \frac{1}{c^2} \, r^2 \, d\Omega^2$$

$$(32)$$

When we cross the Schwarzschild radius, we do have both coefficients changing sign. The signature $- + + +$ is preserved. But something very odd happens: the thing that we are calling t becomes a space-like dimension, and the thing that we are calling r becomes a time-like dimension. They flip.

It is not easy to visualize, but I will later give you the tools to visualize it. Something happens as we cross a certain threshold – the surface of a sphere of radius $r = 2MG/c^2$. The radius r becomes a time variable, and the time t becomes a space variable. This is completely mysterious for the moment, but some diagrams will clarify what is happening at the Schwarzschild radius.

We will discover that, when we are moving radially toward the black hole, to cross the Schwarzschild radius takes an infinite amount of coordinate time t but a finite amount of proper time τ. In other words, somebody with a wristwatch who is falling will say it takes them a finite amount of time to cross that threshold. But somebody standing outside and watching that person fall through will say it takes an infinite amount of time. That is a characteristic of the metric given by equation (32). And that is one of the things we want to work out in this lecture.

Event Horizon of a Black Hole

The peculiar phenomenon of time-space switch is happening at coordinate $r = 2MG/c^2$. The speed of light c is a huge number, and its square is in the denominator. Therefore r is a priori a very small distance. This observation is correct, however, only if M itself is not huge, because the Schwarzschild radius depends on the mass of the gravitating object. For the Earth, if its mass were concentrated in a smaller distance than a centimeter, the Schwarzschild radius would be about a centimeter.

When we cross the Schwarzschild radius, the reader may wonder what happens to his or her wristwatch. Does it begin to record spatial distance? No, it keeps ticking in a time dimension. Clocks don't care about which coordinates we use.

We shall discover that the phenomenon of time-space switch is an artifact of our coordinate system. There is nothing special going on at the Schwarzschild radius. We will see that (t, r, θ, ϕ) are just awkward coordinates: it is they which make it look like something funny is happening at $r = 2GM/c^2$.

Nothing funny is happening at the Schwarzschild radius. We are going to work this out and see exactly why. But for the moment it does look like there is something going on there. When we look at the metric tensor expressed in the coordinates (t, r, θ, ϕ), when the g_{tt} coefficient vanishes, the g_{rr} coefficient becomes infinite. So it looks like terrible things happen to the geometry.

In fact the geometry is completely smooth over the threshold $r = 2GM/c^2$. The light cones are perfectly normal all over the region around the Schwarzschild radius. But we have some work to do to clarify this point.

On a historical note, Einstein submitted his field equations to the Prussian Academy of Science on November 25, 1915. Then Schwarzschild[10] studied Einstein's paper and, in December 1915, while fighting on the Russian front during World War I, wrote down equation (32) for the metric of a gravitating body, when we are outside the body. Unfortunately he died shortly afterward.

What did Einstein already know of the metric? He knew about the first term in front of dt^2, and he knew part of the second term in front of dr^2 in equation (32).

To understand what we mean, let's examine this second term when r is not too small, that is, when $2MG/c^2r$ *is* small. Using our favorite tool to make approximations – the binomial theorem – in equation (32) the coefficient in front of dr^2 can be expanded as follows:

$$\frac{1}{1 - \frac{2MG}{c^2 r}} = 1 + \frac{2MG}{c^2 r} + \left(\frac{2MG}{c^2 r}\right)^2 + \dots \qquad (33)$$

On the right-hand side the first term after 1 has $1/c^2$, the second term $1/c^4$, etc. In other words, what we are doing, in equation (32), is correcting the dr^2/c^2 of equation (30) with correction terms that are very small – of the order of magnitude of $1/c^4$ or smaller. They contribute to the motion of the particles, but not in the non-relativistic limit.

If we are not studying the non-relativistic limit, we don't need to keep the c in our equations. We can choose units in which it is equal to 1. Then the metric looks like this:

$$d\tau^2 = \left(1 - \frac{2MG}{r}\right) dt^2 - \left(\frac{1}{1 - \frac{2MG}{r}}\right) dr^2 - r^2 d\Omega^2 \qquad (34)$$

Two kinds of trajectories are easy to study: circular orbits and radial trajectories. Both will provide some interesting surprises. In this lecture, let's study a radial trajectory: a spacecraft falling along a straight line and going to crash into the black hole.

---

[10]Karl Siegmund Schwarzschild (1873–1916), German astrophysicist. He taught at the University of Göttingen, where he had among his colleagues Minkowski and Hilbert.

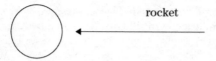

rocket

Figure 10: Radial trajectory of a spacecraft falling toward a black hole.

The question is: how long does it take for the spacecraft to reach the horizon? We are going to measure this duration from two points of view:

1. that of an external observer, using coordinate time t, and

2. that of an astronaut in the spacecraft, using the proper time τ of the spacecraft.

The results are very different.

When we have to do any such calculation, it is best to start with the Lagrangian for the particle and then work out the Euler–Lagrange equations. There are tricks that can be used. I will show some of the tricks that allow us to solve the problem.

Let's go back to the Lagrangian for the in-falling particle or rocket. Remember that particles follow geodesics in space-time, therefore, from equation (11), the Lagrangian is just $-m\sqrt{d\tau^2}$. We get $d\tau^2$ from equation (34). Assuming that the spacecraft goes straight and that Ω doesn't change, we can omit the term $r^2 d\Omega^2$. Thus the action, which must be stationary, is

$$A = -m \int \sqrt{\left(1 - \frac{2MG}{r}\right) dt^2 - \left(\frac{1}{1 - \frac{2MG}{r}}\right) dr^2}$$

As we did before, to make sense out of this integral, below the integral sign we divide, inside the square root, by dt^2, and simultaneously multiply outside the square root by dt:

$$A = -m \int \sqrt{\left(1 - \frac{2MG}{r}\right) - \left(\frac{1}{1 - \frac{2MG}{r}}\right) \frac{dr^2}{dt^2}} \ dt$$

The ratio dr^2/dt^2 is the radial velocity squared, that is $\dot{r}^2$. So we got our Lagrangian:

$$\mathcal{L} = -m \sqrt{\left(1 - \frac{2MG}{r}\right) - \left(\frac{1}{1 - \frac{2MG}{r}}\right)\dot{r}^2} \qquad (35)$$

First of all, there is a conserved quantity, namely the energy.

What is energy in terms of the Lagrangian? Remember, in volume 1 of TTM, the sixth lecture on least action principle. In it we saw that

the energy is the Hamiltonian.

The Hamiltonian itself is expressed in terms of the Lagrangian. To compute the Hamiltonian, the first thing to do is calculate the generalized conjugate momentum of r. It is

$$p = \frac{\partial \mathcal{L}}{\partial \dot{r}}$$

In the general case, when instead of one coordinate r we have a collection of coordinates q_i and their conjugates p_i, the general formula for the Hamiltonian is[11]

$$H = \sum_i p_i \dot{q}_i - \mathcal{L}$$

In our case it becomes simply $H = p\dot{r} - \mathcal{L}$. The calculations are left to the reader. The result is

$$H = \frac{m \ (1 - 2MG/r)}{\sqrt{(1 - 2MG/r) - \dot{r}/(1 - 2MG/r)}} \qquad (36)$$

It is an ugly expression, but it is a definite thing. What does it depend on? It depends on the mass m of the particle or spacecraft, the distance r to the center of the black hole, and the velocity $\dot{r}$. It is the energy. We shall call it E instead of H. It does not change with time.

---

[11]It is equation (4) of lecture 8 of volume 1.

Then equation (36), giving the energy, enables us to express $\dot{r}$ as a function of that energy E. With some algebra we get

$$\dot{r}^2 = \left(1 - \frac{2MG}{r}\right)^2 - \frac{\left(1 - \frac{2MG}{r}\right)^3}{E^2} \tag{37}$$

This expression again looks unwieldy. But it doesn't matter. What is important is that from it we can easily see what happens when r approaches $2MG$, that is, when the spacecraft gets close to the radius (or horizon) of the black hole.

Equation (37) tells us that as the spacecraft approaches the horizon of the black hole, its velocity slows down to zero. Contrary to intuition, when the spacecraft falls toward the Schwarzschild radius, instead of accelerating, its velocity gets smaller and smaller!

Exercise 2: Show that from equation (36) for the energy, and equation (37) for $\dot{r}^2$, it follows that

$$\dot{r} \approx \sqrt{\frac{r - 2MG}{2MG}} \qquad \text{as} \quad r \to 2MG \tag{38}$$

Equation (38) offers another way to see that as r approaches $2MG$, the spacecraft decelerates.

The speed converges asymptotically to zero, and the spacecraft itself never passes the Schwarzschild radius. You might say figuratively that it takes forever to reach the horizon. But that, as we stressed, is in the time frame of an external observer.

Let's go back to the Schwarzschild metric given by equation (34). We write it again leaving aside the $d\Omega^2$ piece, which does not play any role:

$$d\tau^2 = \left(1 - \frac{2MG}{r}\right) dt^2 - \left(\frac{1}{1 - \frac{2MG}{r}}\right) dr^2 \tag{39}$$

We want to examine how a light ray moves. To finish this lecture, we will look at a *radial* light ray.

Motion of a Light Ray

The motion of a radial light ray leads to a special case of equation
(38), which is especially easy to solve. In fact, we don't really
need the Lagrangian equations of motion. All we need to know
is that light rays move along *null trajectories*. That is the name
given to trajectories that satisfy $d\tau^2 = 0$.

Figure 11: Radial light ray beamed from position r in each direction.

Following a null trajectory, a light ray satisfies

$$\left(1 - \frac{2MG}{r}\right)^2 dt^2 = dr^2$$

or equivalently

$$\frac{dr}{dt} = \pm \left(1 - \frac{2MG}{r}\right) \qquad (40)$$

A light ray is of course the fastest thing that can exist; indeed it
moves with the speed of light.[12] When r is big, the right-hand
side of equation (40) is almost 1 in absolute value, which in our
current units is the speed of light.

But as r decreases to the Schwarzschild radius, we see that the
magnitude of the radial velocity, which is $(r - 2MG)/r$, goes to
zero. In other words, even a light ray somehow gets stuck as it
approaches the black hole's horizon!

Is it moving with the speed of light? Yes, of course, it is moving
with the speed of light. What else can a light ray do? But the

---

[12]It is the fastest thing *that we can observe*. We are not talking about
stars very distant from us that can move away from us at speeds higher than
the speed of light.

speed of light c has this property that *in the particular set of co-ordinates (t, r, θ, ϕ), c measured by dr/dt goes to zero as you get closer and closer to the Schwarzschild radius.* Therefore nothing, including light rays, can get past the surface of the horizon. Or so it seems.

The velocity goes to zero at the horizon, whether we are incoming or outgoing. Therefore wherever we are, something odd happens beyond the horizon. We are going to work that out.

Let's turn to someone moving with the rocket of figure 10, or with the photon of figure 11. From the point of view of a person falling in, time is the proper time τ. They will just go sailing right through the Schwarzschild radius. The phenomenon of dr/dt going to zero is an artifact at $r = 2MG/c^2$ of the peculiar spherical polar coordinates used for the stationary frame. There is nothing really going on that is strange at that distance. Moreover the person will reach the black hole in a finite amount of proper time.

We will examine this phenomenon in more detail in lecture 6. In particular, we will see the relation between the stationary time t and the proper time τ. Equation (39) shows that when r approaches the Schwarzschild distance, the coefficient $(1 - 2MG/r)$ goes to zero. When the coefficient is going to zero, a given amount of dt corresponds to a smaller and smaller amount of proper time $d\tau$. As a consequence, proper time in some sense "slows down." That is why something can take an infinite amount of time in one frame (the frame where time is t) and a finite amount of time in the other frame (the frame where time is τ).

One last point we should like to emphasize is that if we were really talking about the Sun or the Earth, these bodies are not nearly compact enough to be black holes. Let's repeat:

A black hole is a body whose mass is contained within its Schwarzschild radius.

The radius is about 3 kilometers for the Sun and about 9 millimeters for the Earth.

Equation (32) or (34) for the metric, or equation (37) for $\dot{r}$, is only valid *outside* the gravitating body itself. Only if the Earth somehow collapsed within a radius of 9 millimeters – while keeping its mass – would there be a Schwarzschild radius to speak about, leading to dr/dt going to zero and so forth when we approach to within 9 millimeters of the center.

The same comment applies to the Sun: only if it collapsed within a sphere less than 3 kilometers in radius would it have a black hole horizon, and strange coordinate artifacts would take place there.

Inside the gravitating body, the metric is different from that expressed by equation (34). And nothing bizarre happens inside the Earth, near its center. There is no black hole horizon of any kind there.

The next lecture will be devoted to deepening our knowledge of black holes.

Lecture 6: Black Holes

Andy: *Lenny, au secours, je tombe dans un trou noir !*

Lenny: *What did you say?*

Andy: *Dépêchez-vous, l'horizon approche !*

Lenny: *Wait a minute, I'll get my French dictionary.*

Andy: *Trop tard, je pense que je l'ai franchi.*

Lenny: *Sorry, I can't hear you.*[1]

Schwarzschild metric
Schwarzschild radius or black hole event horizon
Light ray orbiting a black hole
Photon sphere
Hyperbolic coordinates revisited
Interchange of space and time dimensions at the horizon
Black hole singularity
No escaping from a black hole

Schwarzschild Metric

Last lecture, we began discussing the properties of a black hole and the Schwarzschild metric it creates. Let's review quickly what we did, before going on.

The Schwarzschild solution is an idealized solution in the same sense that the Newtonian force law

$$F = -\frac{mMG}{r^2} \tag{1}$$

is an idealization under the assumption that the mass creating the gravitational field is a point at the origin of the coordinates.

---

[1] Notice that if Lenny stays outside the black hole at some distance from the horizon, in Lenny's proper time Andy's crossing of the horizon happens after an infinite amount of time. So Lenny can *never* hear Andy's last words.

If the mass is spread out, then of course we don't really believe that equation (1) holds in the interior of where there is mass. For instance, in the Earth's interior, equation (1) is not correct. But it is correct in the exterior, beyond the surface of the Earth, at least if we ignore the atmosphere and the other forms of mass.

The point mass of Newtonian mechanics is clearly an idealization because you never have all the mass concentrated in one point. Real materials have a certain stiffness. You can compress them but you can never squeeze them to an infinitely small radius. They just have a certain resistance to being squeezed.

Gravity is only so strong, even though the gravitational field appears to get very, very large at short distances. The nature of all materials is always such that if you squeezed them down to arbitrarily small distance, they will spring back.

These natural considerations, however, do not transfer without some profound changes to general relativity. We are going to see an example where, if matter gets too close to the singularity, which is itself by definition at $r = 0$, it gets sucked in in a way that doesn't happen in Newtonian gravity.

To take another illustration of the differences between Newtonian physics and general relativity, in Newtonian physics, if we have a force center, and we shoot a particle with infinite precision along a radius toward the force center, it will indeed hit the center. But if we deviate in the slightest way from a perfectly radial shot – and that is true even if the momentum mv given to the particle is not particularly big – the particle will not hit the center. It will swing around it as shown in figure 1.

There is a kind of effective force repelling the particle from the center.[2] It is a centrifugal force. If you don't aim perfectly, the particle will have some angular momentum relative to the center, and the effective centrifugal force will keep it from going to the center.

---

[2]Remember that in the physical context here the word *effective* is synonymous with *fictitious*.

In Newtonian mechanics, even though the center pulls everything very hard toward itself, when you get close to it, it is nevertheless infinitely unlikely for an in-falling point particle to actually hit the center.

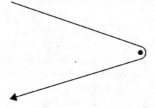

Figure 1: Shooting a particle toward a force center without perfect precision. In Newtonian mechanics, the resulting trajectory is a branch of a hyperbola that, viewed from afar, looks almost like two straight half lines.

The situation is quite different in the general theory of relativity. Gravity is even stronger. It is so strong that it overwhelms the centrifugal barrier and pulls anything into the center.

Let's go back to the Schwarzschild metric that we began to study in lecture 5. Notice that we did not derive it. We just wrote it with various heuristic justifications. We need Einstein's field equations, which will be established in lecture 9, to show that the Schwarzschild metric is a solution to them. So far for us it is a specific metric, i.e., a specific geometry, that is a given. Our objective presently is to analyze the consequences of the metric. The metric is expressed (with $c = 1$) by the following formula:

$$d\tau^2 = \left(1 - \frac{2MG}{r}\right) dt^2 - \left(\frac{1}{1 - \frac{2MG}{r}}\right) dr^2 - r^2 d\Omega^2 \quad (2)$$

or equivalently

$$dS^2 = -\left(1 - \frac{2MG}{r}\right) dt^2 + \left(\frac{1}{1 - \frac{2MG}{r}}\right) dr^2 + r^2 d\Omega^2 \quad (2')$$

To see where c appears, go to equation (32) of lecture 5.

Remember that $d\Omega^2$ is simply the ordinary metric on the unit sphere expressed with two spherical angles, the angle θ from the equator and the azimuthal angle ϕ; see figure 9 and equation (29) of lecture 5. The metric on the unit sphere in spherical coordinates, therefore with $r = 1$, is expressed as

$$d\Omega^2 = d\theta^2 + cos^2\theta \ d\phi^2 \qquad (3)$$

Remember also that as r crosses the value $2MG$, in equation (2) the coefficient in front of dt^2 changes sign, but so does the coefficient in front of dr^2, therefore the signature of the metric dS^2 (which is simply the opposite of $d\tau^2$ when we work with $c = 1$) retains essentially the same structure.

Regarding the two spherical angles, θ and ϕ, sometimes we may be interested in only one of them. This is the case for circular orbits. For instance, when we study an orbiting particle, we can locate it in space-time with t and ϕ. Radius r will remain constant and $\theta = 0$ – if we use latitude measured from the equator as in figure 9 of lecture 5. If we used the polar angle – i.e., position measured from the pole – we would have $\theta = \pi/2$.

Schwarzschild Radius
or Black Hole Event Horizon

Let's draw a picture of the metric, with the time axis t vertical, and two spatial axes x and y forming a horizontal plane at $t = 0$; see figure 2. Of course, there should be three spatial axes, x, y and z, and the plane at $t = 0$ should really be a 3D space, but it is not possible to draw such a thing, together with the time axis, on a page.

The radius $r = 2MG$, corresponding to the cylinder,[3] is called the *Schwarzschild radius*, or the *event horizon* of the black hole. We will understand why as we go along.

Let's look at what is going on very far away. When r is very big, the coefficient in front of dt^2, in equation (2), becomes almost 1,

---

[3]We will usually call the circle that the cylinder cuts in the (x, y) plane in figure 2 a sphere, because we think of three spatial coordinates.

and the coefficient in front of dr^2 almost -1. Then it is just flat space-time in polar coordinates.

As we move along decreasing r and eventually cross the sphere defined by $r = 2MG$, something happens. The geometry becomes very different. Something looks bad at that value of r. First of all, the coefficient of dt^2 becomes zero. But, even worse, the coefficient of dr^2 becomes infinite. Our equation somehow blows up.

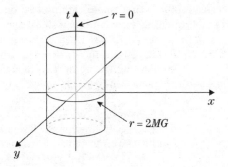

Figure 2: Space-time in the vicinity of a black hole.

However we will see that nothing really singular happens at $r = 2MG$. The "blow up" is a coordinate artifact that can be removed by a new choice of coordinates.

At $r = 0$ there is also a problem in the equation. And this time it does correspond to a real singularity in the physics. The coefficient in front of dt^2 becomes infinite, and that in front of dr^2 zero. What is going on at $r = 0$ is that the curvature is becoming infinite there.

At the Schwarzschild radius $r = 2MG$, no large curvature or enormous deviation from flat space is observed. But at $r = 0$ all hell breaks loose. If we calculated the curvature tensor to find out how curved the geometry is at the center, we would find that all of its components become infinite.

There is another way to describe that phenomenon. Curvature means tidal forces. Anything falling in, when it would hit $r = 0$, would experience infinitely strong tidal forces and be torn to shreds. It would be stretched radially and squeezed between two angular directions by an infinite amount. So $r = 0$ is the actual place where disaster occurs. It is not at the event horizon.

We talked about radially in-falling things. Let's go over the conclusion we reached. We won't need to go into the mathematics again, because this time we are going to see what is happening with the help of pictures. We will see again that from one point of view it takes an infinite time for something to cross $r = 2MG$, while from another point of view it takes a short time.

In figure 2, the surfaces of constant time t are the horizontal planes at different heights. Something falling in, say, along the x-axis, would have a trajectory shown in figure 3.

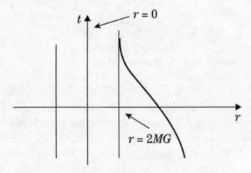

Figure 3: Trajectory of a particle falling toward a black hole, in spherical polar coordinates (with $\theta = \phi = 0$).

The particle or object accelerates for a while. But as it gets closer and closer to the horizon, its trajectory approaches asymptotically the vertical axis $r = 2MG$, while it never gets quite there. If the object is a meter stick falling in, with its direction pointing to the center, i.e., oriented radially, its front end and back end would follow the trajectories shown in figure 4.

The only way to make sense out of this phenomenon is to say, *at least in these coordinates*, that the gap in r between the front end and the back end of the meter stick shrinks to zero.

It is actually a form of Lorentz contraction. As something falls in, even though it appears to be slowing down, its momentum in fact is increasing, and the object is Lorentz-contracting.

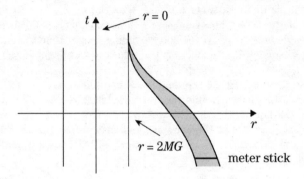

Figure 4: Meter stick falling in radially.

Think next about a clock. Let's imagine that the meter stick was a clock. How do we build a clock that is also a meter stick? We can use two mirrors, one at each extremity of the meter stick, and allow a light beam to go back and forth, and count the number of reflections. But clocks as they move – whatever motion they have – don't tick off time t of equation (2). They tick off proper time. It is the meaning of the proper time: it is the ticking of a clock traveling together with the moving object under study.

Time t of equation (2) is called *coordinate time*. It is just the coordinate we are using to describe the entire system. Remember that space-time exists irrespective of the coordinate system we use to chart it. We can use many different ones. The Schwarzschild metric is expressed in equation (2) with the spherical polar coordinates of a stationary observer sitting outside the black hole.

Notice that if we are very far away, where r is very big, and if we are not moving, then $d\tau$ is equal to dt, at least to a very good approximation, because $1 - 2MG/r$ is essentially 1. And dr and $d\Omega$ are 0 for someone who doesn't move. In other words, very far away, proper time and ordinary time are the same.

As we get closer and closer to the black hole, though, for a given amount of dt, we see from equation (2) that the incremental proper time $d\tau$ gets smaller and smaller. For instance, consider the coordinate time elapsing from t_1 to t_2. Think of t_1 and t_2 as horizontal lines in figure 4. How much proper time elapses *along the trajectory* of the object between these two horizontal lines? When t_2 increases by dt over t_1, the corresponding proper time increment $d\tau$ gets smaller and smaller, as the values t_1 and t_2 get bigger.

While it appears that the amount of coordinate time t it takes to reach the Schwarzschild radius of the black hole is infinite, we could ask how much τ has elapsed by the time an in-falling object eventually arrives at the horizon. And that proper time is finite. Again we are going to see this in pictures, without any mathematics.

Consider somebody falling in with a clock. Will that person experience an infinite amount of time before they get to the horizon, or a finite amount? (By the way, the clock doesn't have to be a mechanical device. By "experience" we mean the time the person feels. It can be heart beats, or any other physiological process that is also a time-keeper.) The answer is: they experience the proper time. So the question is: how much proper time does it take to reach the horizon?

To answer the question, we could solve equations, we could calculate, do integrals. But fortunately it is not necessary. Diagrams will directly give the answer.

Before we get to the diagrams, let's solve another problem that does not have to do with somebody falling into a black hole, but with another phenomenon: orbiting the black hole.

Light Ray Orbiting a Black Hole

Orbiting a black hole means going round it as time t goes on. When we think about a circular orbit, in the space-time representation of figure 2, the trajectory will have the shape of a corkscrew (either an ordinary "right" corkscrew or a "left" corkscrew).

If we don't need to represent the coordinate time t-axis, we can figure out the trajectory as in figure 5. It is a view showing only space. Think of figure 2 "viewed from above."

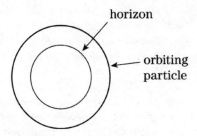

Figure 5: Particle orbiting a black hole.

The center of the black hole has coordinate $r = 0$, and the horizon $r = 2MG$. The radius at which the particle is orbiting is some larger value.

In fact we won't study the trajectory of any kind of particle; we will study the trajectory of a light ray. As we know from Einstein's starting point in his work on general relativity, a light ray passing by a massive body is bent.

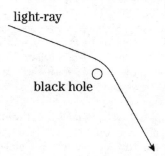

Figure 6: Light ray trajectory bent by a black hole (or any other massive body).

In lecture 1 we saw that in Newtonian physics a uniformly accelerated elevator has an effective gravitational field inside it. From that, we inferred the *equivalence principle*: accelerated frames and gravitation fields are equivalent.[4]

Secondly, we saw that a light ray crossing the elevator appears bent to someone in the elevator, whence we deduced that real gravity should bend light rays. Then in lecture 4, we saw that in special relativity a uniformly accelerated frame of reference – which was something subtle to define – shows an effective gravitational field.

In figure 6 the same thing is happening: gravity bends trajectories. And obviously as the light ray passes closer to the black hole, it will be bent more. Why? Because gravity becomes stronger. As it skims the Schwarzschild radius closer and closer, the light ray really whips around the black hole. Eventually, at some point, not at the horizon but a bit farther out, the trajectory will circle the object in a closed orbit.

By the way, this will never happen near the surface of the Earth. You would have to squeeze the Earth to a small enough volume to transform it into a black hole. We already said that this would be a ball – with the same mass as the entire Earth! – of radius less than 9 millimeters. So we are not going to see light rays circling around the Earth. Nevertheless it is very illuminating to try to understand orbiting light rays around black holes.

The next step is somewhat calculation-intensive. I'll show the sequence of operations to calculate the circular orbits. We are going to do it using the rules of classical mechanics that we learned in volume 1. The reason we will do it this way is because it is the only easy way, and even then it is not so easy.

An important point to keep in mind is that the material we learned in classical mechanics is not disconnected from the rest of physics and from what we are doing now. The principle of least action is universal. It applies in classical mechanics, quantum mechanics,

---

[4]Recall that in the accelerated elevator we felt only an apparent effect, also called effective phenomenon. It can be compared only to a special, ideal gravitational field because it is a field with no curvature, no tidal effect. This is approximately true in any very small region of a real gravitational field created by a massive body.

special and general relativity, Maxwell electrodynamics, quantum field theory, etc. The rules for deriving the Hamiltonian from the Lagrangian remain the same.

We shall use some conservation laws, plus the basic principle of equilibrium, which says this: equilibrium happens at stationary points of potential energy. Finally we are going to use the mechanics of an object orbiting, but not an object moving in space-time according to Newtonian mechanics, an object moving in space-time with the Schwarzschild metric given by equation (2).

We already talked about it. Yet let's review quickly the principles of the Euler–Lagrange minimization approach. Then we will go into the calculations. We start with the action. Whenever you are working out equations of motion for anything, it is almost always easiest to start with the action principle. From the action principle derive a Lagrangian, and from the Lagrangian derive the Euler–Lagrange equations of motion. Or, from the action, use some other trick of the same nature.

In our case we are going to use conservation laws. We start with the metric given by equation (2), written again here:

$$d\tau^2 = \left(1 - \frac{2MG}{r}\right) dt^2 - \left(\frac{1}{1 - \frac{2MG}{r}}\right) dr^2 - r^2 d\Omega^2$$

In units with $c = 1$, the action of a moving particle[5] of mass m is

$$A = -m \int d\tau \qquad (4)$$

The proper time increment $d\tau$ is the square root of the right-hand side of the metric equation defining $d\tau^2$. Let's write the action more explicitly. In order not to write over and over the coefficient $(1 - 2MG/r)$, we will denote it $\mathcal{F}$, and its inverse $\mathcal{F}^{-1}$.

With these notations, the action of equation (4) is written

$$A = -m \int \sqrt{\mathcal{F}dt^2 - \mathcal{F}^{-1}dr^2 - r^2 d\Omega^2} \qquad (5)$$

---

[5] We begin the calculations with a massive particle. Then we will pass to the limit when $m \to 0$ in order to study massless photons.

We are considering a particle circling around the black hole, so in the coordinates $(t, \ r, \ \theta, \ \phi)$, we can take one of the angles to be fixed and the other to be the angular parameter of the orbit. Going back to figure 9 of lecture 5, suppose that the angle θ, which we chose to measure from the equator, is fixed at zero, then $d\theta = 0$. The azimuthal angle ϕ, which varies from 0 to 2π, will be used to track the orbiting particle. From equation (3), we deduce that $d\Omega^2 = d\phi^2$.

By now the reader should be familiar with the next step: beneath the integral, we divide by dt^2 inside the square root and multiply by dt outside the square root. Equation (5) becomes

$$A = -m \int \sqrt{\mathcal{F}(r) - \mathcal{F}^{-1}(r)\dot{r}^2 - r^2\dot{\phi}^2} \ \ dt \qquad (6)$$

For a circular orbit, $\dot{r} = 0$. Therefore the formula can be simplified further, but we don't want to quite go there yet.

What do we read off the Lagrangian? Remember that the Lagrangian is the quantity under the integral sign when $-m$ has been brought back in. The factor dt is not part of the Lagrangian; it is the infinitesimal time multiplying it. As usual we denote the Lagrangian by the letter $\mathcal{L}$:

$$\mathcal{L} = -m \ \sqrt{\mathcal{F}(r) - \mathcal{F}^{-1}(r)\dot{r}^2 - r^2\dot{\phi}^2} \qquad (7)$$

There are two conservation laws we want to use.

1. **Conservation of energy**. When the Lagrangian of a system doesn't explicitly depend on t, the energy is conserved.

2. **Conservation of angular momentum**. The angular momentum of an orbiting particle is the conjugate momentum to ϕ. When the Lagrangian is invariant under rotation, it is also conserved.[6]

---

[6]The angular momentum is the analog for rotating objects of the linear momentum – simply called *momentum* or *impulsion* – for objects moving in a straight line. Remember too that, if $\mathcal{L}$ is invariant under translation, the momentum of the object, or system of particles, is conserved. These results are various applications of Noether theorem; see volume 1 of TTM.

Let me remind you of the definition of conjugate momentum in Lagrangian mechanics: if we have a coordinate q – also called a *degree of freedom* – the momentum associated with q is equal to the derivative of the Lagrangian with respect to $\dot{q}$. We have seen and used it repeatedly in volume 1. We use it again here. Our degree of freedom is the angle ϕ. Let's call L its conjugate momentum. It is naturally called the *angular momentum*:

$$L = \frac{\partial \mathcal{L}}{\partial \dot{\phi}} \tag{8}$$

Using the expression for $\mathcal{L}$ given by equation (7) and the chain rule of differentiation, we first write the derivative of the square root of the intermediate variable g with respect to g, where

$$g = \mathcal{F}(r) - \mathcal{F}^{-1}(r)\dot{r}^2 - r^2\dot{\phi}^2 \tag{9}$$

Then we multiply it by the derivative of g with respect to $\dot{\phi}$, carrying along the factor $-m$. Two factors 2 and the minus signs disappear. We get

$$L = \frac{mr^2\dot{\phi}}{\sqrt{g}} \tag{10}$$

Since the Lagrangian $\mathcal{L}$ doesn't depend on ϕ – only on $\dot{\phi}$ – the angular momentum L is constant. It is also proportional to the mass m of the particle.

Let's rewrite the angular momentum more explicitly:

$$L = \frac{m\,r^2\dot{\phi}}{\sqrt{\mathcal{F}(r) - \mathcal{F}^{-1}(r)\dot{r}^2 - r^2\dot{\phi}^2}} \tag{11}$$

As said, for a circular orbit it simplifies a bit because $\dot{r}$ is equal to 0. But we are not doing the circular orbit yet.

Note that L was the conjugate momentum associated with ϕ. Let's turn our attention to the other degree of freedom, the radius r, which also has a conjugate momentum. We denote it P_r. To find P_r, we also have to differentiate the Lagrangian, this time with respect to $\dot{r}$. We find

$$P_r = \frac{\partial \mathcal{L}}{\partial \dot{r}} = \frac{m\mathcal{F}^{-1}\dot{r}}{\sqrt{g}} \tag{12}$$

Unlike the angular momentum L, which was conserved because the Lagrangian did not depend on ϕ, the *radial momentum* P_r is not conserved. Indeed, the Lagrangian depends on r. As a consequence there is a radial potential, and there are radial forces. We could calculate them from the Euler–Lagrange equations. They determine accelerations along the radial axis.

If there are accelerations along the radial axis, we see again that P_r cannot be conserved. But what is conserved is the energy. Let's write the general expression for the energy.

It is of course the good old Hamiltonian. Remember the expression of the Hamiltonian (see volume 1, lecture 8, equation (4)). If we have N generalized coordinates q_i, the index i running from 1 to N, and the p_i's are the conjugate momenta, and $\mathcal{L}$ is the Lagrangian, then the Hamiltonian H is

$$H = \sum_i p_i \dot{q}_i - \mathcal{L} \qquad (13)$$

The point is that there is always an energy given by equation (13). Equation (13) actually *defines* it. And the energy is conserved so long as the Lagrangian $\mathcal{L}$ does not explicitly depend on time.

In our case, there are only two coordinates, r and ϕ. From equation (13), we can calculate H. It is

$$H = \frac{\mathcal{F}(r)\,m}{\sqrt{g}} \qquad (14)$$

The letter H refers to the name of William Hamilton (1805–1865). In what follows however, we shall denote the energy with the letter E.

$$E = H$$

Let's re-express the energy given by equation (14), but now we shall write explicitly the quantity that we denoted g. And we are going to use the fact that we are on a circular orbit. Hence the radius doesn't change: $\dot{r} = 0$. The energy takes then the form

$$E = \frac{\mathcal{F}(r)\,m}{\sqrt{\mathcal{F}(r) - r^2 \dot{\phi}^2}} \qquad (15)$$

This is partly kinetic energy due to the motion of ϕ. Note that there is no kinetic energy due to the motion of r, because r is not changing on a circular orbit. And it is partly potential energy.

The energy E combines potential energy, because it depends on r, and kinetic energy, because it also depends on the angular velocity $\dot\phi$. The important point is that it is conserved.

What about the angular momentum? We got it in equation (11). It has the same denominator as the energy. We can write it now

$$L = \frac{m\,r^2\dot\phi}{\sqrt{\mathcal{F}(r) - r^2\dot\phi^2}} \tag{16}$$

It is also conserved.

What do we do with these two conserved quantities given by equations (15) and (16)? First of all, we solve equation (16) for $\dot\phi$. This will give us the angular velocity as a function of the angular momentum. It is very easy and is left to the reader. Then we plug it back into equation (15) for the energy. The result will be an expression of the energy E as a function of r and the angular momentum. We will write in a moment what we get.

Physics always consists of this alternation between principles – having fun figuring out the principles, and what they tell us to do – and the boring work of calculations. This second part is like turning the crank – not quite number crunching but "letter crunching," if you will. Then we go back to the principles and see what the results say in light of the principles.

Let's shortcut the calculations. Here is what we get for the energy:

$$E = m\,\frac{\sqrt{\mathcal{F}(r)}\,\sqrt{r^2\left(\frac{L}{m}\right)^2 + r^4}}{r^2} \tag{17}$$

This does not depend on $\dot r$. It depends only on r and L. What we are interested in is the value of r that corresponds to a position of equilibrium of the energy. The equilibrium position depends – as intuition suggests – on the angular momentum.

In the preceding analysis, we considered a particle with some mass m. Now we want to go to the limit of a photon orbiting the black hole on a circular trajectory. A photon is massless, so a priori the analysis should collapse. However it can easily be adapted.

Photons also have momentum and angular momentum. A photon moving around in orbit, or even if it is just going past the star, does have an angular momentum with respect to the star.

In equation (17), take the factor m that is in front, and introduce it inside the second square root. Inside that square root we get

$$r^2 L^2 + m^2 r^4 \tag{18}$$

The angular momentum L is constant, and the term $m^2 r^4$ goes to zero. After simplification, we end up with the simple formula

$$E = L \, \frac{\sqrt{\mathcal{F}}}{r} \tag{19}$$

This shows that, for a given angular momentum L – which is just a multiplicative factor – the energy of the photon depends only on r. You can think of this E as a kind of potential energy.

The radial velocity $\dot{r}$ has been eliminated since on a circular orbit there is no radial velocity. The angular velocity $\dot{\phi}$ has been eliminated by expressing it in terms of the angular momentum L. Thus the energy E of the photon depends only on r. Let's now study it. In particular, let's find its point of equilibrium.

Photon Sphere

The question we are interested in is: where is the equilibrium of E in terms of r? To find the value of r, all we have to do is differentiate the right-hand side of equation (19) with respect to r and set the derivative equal to zero. At that value of r, the function $L\sqrt{\mathcal{F}(r)}/r$ will either have a maximum or minimum value. Remember that the Schwarzschild factor, which we decided to denote $\mathcal{F}(r)$ to make the expressions lighter, is

$$\mathcal{F}(r) = 1 - \frac{2MG}{r}$$

When r becomes very big, E is asymptotically equal to L/r. And, on the contrary, when r gets close to the horizon $2MG$, E gets close to zero. In between E is positive, and there is a point where its derivative is zero.

In figure 7 is plotted the energy, given by equation (19), as a function of r. We see clearly that it reaches a maximum before declining as r increases. Elementary calculations left to the reader show that the point where E is stationary is at

$$r = 3MG \qquad (20)$$

It does not depend on the angular momentum. The interesting fact is that where E is stationary, it is at a maximum.

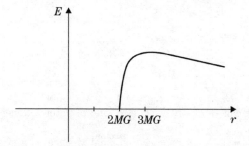

Figure 7: Energy E of a photon as a function of r.

If you have a potential energy that has a maximum, does that correspond to an equilibrium position? The answer is yes, but it is an unstable equilibrium. It is like placing a marble on top of a smooth metal sphere: if you place it absolutely precisely, the marble will stay there. But the slightest misplacement off the top, or the slightest flick, and the marble will roll off. The smaller the flick you give it, or the smaller the imprecision in putting it right up at the top, the longer it will last at or near the top before eventually rolling down.

Here we are not talking about marbles on smooth spheres, but about the circular orbit of a massless particle around a black hole. What we found is that exactly at $r = 3MG$ a photon can orbit a black hole. It is not at the Schwarzschild radius. It is at one and a half times that distance. For obvious reasons, the sphere at that radius is called the *photon sphere*.

You might expect that if you had a black hole, and you looked at that distance, you would find photons orbiting around it. After all any photon that gets started in that orbit, will stay in that orbit.

Of course, if the photon starts slightly away from the photon sphere at $r > 3MG$, it will whip around and come out – not necessarily in the opposite direction. The closer you get to the photon sphere, the more angle the photon will sweep before it eventually goes away. A photon coming in very close can circle around several times before escaping; see figure 8.

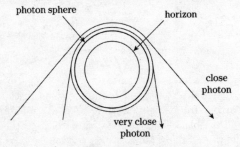

Figure 8: Photons passing slightly outside the photon sphere.

Photons whipping around or temporarily orbiting the black hole before escaping correspond to a distance to the black hole slightly off to the right of $3MG$ in figure 7, in other words, slightly farther than the radius of the photon sphere.

What happens, on the other hand, if a photon is on the left of $3MG$? It will get sucked in. If it finds itself slightly inside the photon sphere, it can't withstand gravity. It might orbit a few times, but it will spiral first to the horizon and then of course to the singularity.

What we just did was an example of calculations with the Schwarzschild metric. We showed the surprising fact that massless particles can have an unstable equilibrium orbit at radius $r = 3MG$.

We recommend that the reader compare the calculations we did with the corresponding Newtonian calculations. You will see that all the pieces are the same. But the outcome is quite different.

In Newtonian physics, light rays don't orbit anything. In inertial frames of reference, they move in straight lines whatever the field. Massive particles in a central gravitational field move along conics. And with the right configurations of E and L, massive particles can follow a circular orbit around a force center, not only at distance $r = 3MG$, but *at any distance*.

Whereas, in space-time with the Schwarzschild metric, particles below the photon sphere, whether massive or massless, will inescapably get pulled in.

The phenomenon of photons temporarily orbiting near the photon sphere before escaping again has interesting consequences.

Let's look at a black hole from a distance. The black hole itself emits no light, but light rays coming from outer space behind the black hole, relative to you, display strange patterns. For instance, a light ray arrives from a light emitter, say, a distant star, in outer space behind the black hole. It may bend a bit and then hit your eye, as in figure 9. So instead of seeing the light ray coming from its real direction in the backdrop, you see it apparently coming from another direction.

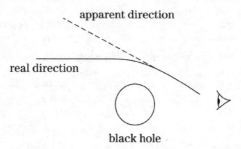

Figure 9: Distant light emitter apparently displaced.

It is worse than that. A light ray coming from the same distant source may pass the black hole near another point, for instance

not above but below the black hole of figure 9. Then it will appear
to be coming from yet another direction than the first one. The
result is that the light emitter behind the black hole will appear
to your eye as a ring around the black hole. A black hole is a bad
place for your enemy to hide from you.

What you see is even worse. Other points in the backdrop may
emit light rays that will come near the photon sphere, orbit for a
while, then happen to leave in a direction that will hit your eye,
as in figure 10.

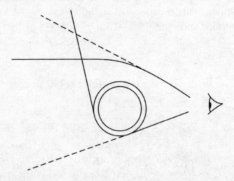

Figure 10: Another light emitter apparently displaced.

Light emitters don't have to be far away to display this phenomenon. Light-emitting points near the horizon of the black
hole, anywhere on the sphere, can emit light rays that will come
out of the photon sphere. Some can also end up in your eye.

In short, what you will see from the backdrop as well as from the
vicinity of the black hole will form a complicated pattern in your
eye due to light ray trajectories modified in various ways by the
black hole. That makes looking at a black hole a rather dizzying
experience.

Notice that the ring created by a source behind the black hole,[7]
called an *Einstein ring*, is a physically observable phenomenon.

---

[7]Black holes are convenient massive bodies to observe light deviation,
because they don't emit light on their own. Therefore, unlike with the Sun,
we don't have to wait for an eclipse to study light rays flying in their vicinity.
See also the topic of gravitational lens.

The first complete one was observed by the Hubble space telescope in 1998.

In recent years, supposedly sensational "photographs" of black holes have been circulating in the popular science press. As the reader understands, what they show is somewhat different and more complicated than what the articles seem to want us to think.

Exercise 1: Explain why a light ray emitted from inside the photon sphere can escape, but a light ray cannot enter the photon sphere and come out again.

Hyperbolic Coordinates Revisited

Now we want to understand better what is happening at the event horizon of the black hole; see figure 11. As we pointed out, something curious happens in the equation of the metric, equation (2) or its variant (2′). The sign of the coefficient in front of dt^2 changes. At the same time the sign of the coefficient in front of dr^2 changes too – thus preserving the signature of the metric.

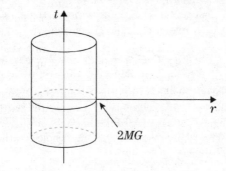

Figure 11: Black hole horizon.

We know that clocks "slow down," in the sense that when r gets close to $2MG$, for a given dt the corresponding $d\tau$ is smaller and smaller. We saw that a stick falling in vertically contracts, etc.

But I also explained that in some other sense, nothing is happening. Someone falling in will not experience anything special while crossing the horizon.

We will pursue our study with the following idea in mind: *many things that happen in gravity can be understood by first understanding them for accelerated coordinate frames in a flat space-time.* It is the meaning and the usefulness of the equivalence principle. Gravity is not exactly the same as a uniformly accelerated coordinate system of course, but it is often a good idea, when studying a question, to first look at how it manifests in a uniformly accelerated coordinate system. So let's go back to that.

In lecture 4, we saw that defining a uniformly accelerated reference frame in relativity is more complicated than in Newtonian physics. Furthermore it is best studied in a coordinate system which is the analog of polar coordinates, but is *hyperbolic* polar coordinates. Let's review it quickly.

We are in a flat space-time now. It is not a black hole, nor the gravitational field of a massive body, it is plain old flat space-time. We have a usual frame in which any point P has standard Minkowski coordinates denoted $(T,\ X)$. In another frame, it has hyperbolic polar coordinates denoted $(\omega,\ \rho)$. There may be more spatial coordinates Y and Z, but we leave them aside because they don't play a role in the subsequent analysis.

We called coordinate ω the *hyperbolic angle*, because it is the analog of the ordinary angle in polar coordinates. However, instead of varying from 0 to 2π, it varies from $-\infty$ to $+\infty$. It is a kind of time. As regards the other coordinate, ρ is sometimes called the *hyperbolic radial coordinate*. A point, i.e., an event, in space-time can be located with its T- and X-coordinates, and also with its ω- and ρ-coordinates; see figure 12.[8]

---

[8]Figure 12 is the standard Minkowski diagram of flat space-time in special relativity, with two axes, the horizontal 1D spatial axis, and the vertical time axis. The diagram also enables us to show the construction of *any other* frame of reference, that is, any other "labeling system" of all the events. In the Minkowski diagram the trajectory of a particle moving with constant velocity on the X-axis is a straight line with a slope equal to or greater than 45°. The diagram also nicely shows the trajectories of points uniformly accelerating.

Remember the following facts on figure 12:
- points on a horizontal line have the same T
- points on a vertical line have the same X
- points on the same dotted line have the same ω
- points on the same hyperbola have the same ρ.

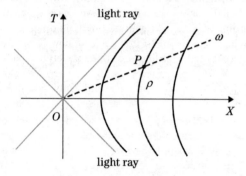

Figure 12: Flat space-time with standard Minkowski coordinates (X, T). Hyperbolas of constant acceleration are shown together with their hyperbolic polar coordinates (ρ, ω).[9]

Finally the algebraic correspondence between the Minkowski coordinates and the hyperbolic polar coordinates of P is

$$X = \rho \cosh \omega$$
$$T = \rho \sinh \omega \qquad (21)$$

What about the Minkowski squared distance $X^2 - T^2$?

In hyperbolic polar coordinates, it becomes $\rho^2(\cosh^2 \omega - \sinh^2 \omega)$, which is equal to ρ^2. So on each hyperbola in figure 12, $X^2 - T^2$ is constant when ω varies.

---

[9]From figure 12 on, when we use the standard Minkowski diagram of flat space-time, the Minkowski coordinates are denoted with the capital letters X and T. In particular, it will be the reference frame of an observer free-falling through the horizon of the black hole, and describing what is seen as in flat space-time. They will see objects fixed in space as accelerating.

We studied transformation (21) in lecture 4. It enabled us to define a *uniformly accelerated reference frame in special relativity*. Let's recall the reasoning:

1. In Newtonian physics:

   (a) The concept of a uniformly accelerated collection of points, fixed relative to each other, is intuitive and clear.

   (b) Points rotating on a given circle, all fixed relative to each other, are all subjected to the same acceleration (in magnitude) toward the center of the circle.

2. In special relativity:

   (a) A collection of points, each with its own fixed ρ, and all moving with the same ω (see the dotted line in figure 12 going through O and whose angle increases up to 45°), are said to be uniformly accelerated. In this latter case, it is a definition.

   (b) Just like on concentric circles, points with the same angular velocity have decreasing acceleration as the radius increases, on different hyperbolas points moving with the same ω have different accelerations.

The metric of ordinary flat space-time, expressed in the hyperbolic polar coordinates, is

$$d\tau^2 = \rho^2 d\omega^2 - d\rho^2 \qquad (22)$$

It is the analog of $r^2 d\phi^2 + dr^2$ in ordinary polar coordinates on the Euclidean plane, i.e., in ordinary geometry. Equation (22) is its equivalent in Minkowski flat space-time of special relativity.

In order to study conveniently what is going on in the vicinity of the event horizon of a black hole, we are going to change coordinates. We don't change coordinates just for the heck of it, but in order to make it easy to compare with the Schwarzschild metric. The new coordinates will be denoted ξ (the Greek letter pronounced "ksi") and t. They are defined as follows:

$$\xi = \frac{\rho^2}{4}$$

$$t = 2\omega \tag{23}$$

Now let's re-express the metric. Elementary algebraic manipulations on equation (22) yield

$$d\tau^2 = \xi dt^2 - \frac{d\xi^2}{\xi} \tag{24}$$

So far, this is just the ordinary Minkowski metric equation expressing the square of the increment of proper time in flat spacetime. But it is expressed in coordinates ξ and t.

We notice a similarity with the Schwarzschild metric. In the Schwarzschild metric, equation (2) or its variant $(2')$,[10] we have a coefficient in front of dt^2, and we have the inverse of the same coefficient in front of dr^2. If we think of ξ as r, equation (24) bears some resemblance.

Notice another point: $(1 - 2MG/r)$ and its inverse both simultaneously change sign when r crosses the horizon. Somehow analogously, ξ changes sign simply when it becomes negative, and so does $1/\xi$ simultaneously of course. Does ξ becoming negative means something? Yes, we shall see what it means.

Now we shall go the other way around. We shall start from the Schwarzschild metric and produce coordinates such that the metric, in some local region, is similar to (in fact exactly like) equation (24); but it will be an approximation.

On the black hole, we want to focus on the place where $r = 2MG$, that is, we want to study the Schwarzschild metric of equation (2) in a small region near the horizon.

The first coefficient in the Schwarzschild metric is $(1 - 2MG/r)$,

---

[10]In what follows we choose to work with the proper time. We won't repeat each time that we could equivalently work with the proper distance.

which can be rewritten

$$\frac{r - 2MG}{r}$$

Since we won't let r change very much near $2MG$, to the first order approximation (in $\epsilon = r - 2MG$) it is equal to

$$\frac{r - 2MG}{2MG}$$

This is like in Newton's mechanics: if you study a phenomenon, a movement near the surface of the Earth, for many purposes you can set r equal to the radius of the Earth.

Omitting for the moment the term $r^2 d\Omega^2$, which only concerns changes in the spherical angles, the Schwarzschild metric becomes

$$d\tau^2 = \left(\frac{r - 2MG}{2MG}\right) dt^2 - \left(\frac{2MG}{r - 2MG}\right) dr^2 \qquad (25)$$

To simplify the expression further, we set the mass of the black hole to a specific number in such a way that $2MG = 1$. We can always do that without loss of generality. It is just a choice of units.

What are the units of $2MG$? They are units of length, because $2MG$ is the Schwarzschild radius.[11] For a particular black hole we can choose units of length where the Schwarzschild radius is just one. We cannot do it for all black holes simultaneously, but if we are interested in one particular black hole, it is not a problem. That makes formula (25) much simpler. It now becomes:

$$d\tau^2 = (r - 1)dt^2 - \left(\frac{1}{r - 1}\right) dr^2 \qquad (26)$$

Next step is to redefine $r - 1$: we call it ξ. It is just a change of variables. Since with our preceding change of units, the horizon had coordinate $r = 1$, we are now translating coordinates so that the horizon has coordinate $\xi = 0$. In other words ξ measures the deviation from the horizon. Finally equation (26) becomes

---

[11]The implicit presence of the factor $c^2 = 1$ in equation (25) makes it dimensionally consistent; see equations (1), (1'), (2), and (2') of lecture 5.

$$d\tau^2 = \xi dt^2 - \frac{1}{\xi}dr^2 \qquad (27)$$

Notice that dr and $d\xi$ are the same thing since $\xi = r - 1$. After all these various changes (local analysis at the horizon, change of units of length, change of units of mass of the black hole, and change of spatial variable near the horizon), the Schwarzschild metric is now

$$d\tau^2 = \xi dt^2 - \frac{d\xi^2}{\xi} \qquad (28)$$

This equation (28) is exactly the same as equation (24). It tells us that the vicinity of the horizon is, in some at-first-sight startling way, just *flat space-time*, or approximately flat space-time. Nothing dramatic in the geometry is happening there. No large curvature, only more or less flat space-time. This may be surprising because of what we saw earlier – the slowing down of clocks, the contraction of lengths, etc. – but as stressed *these were just coordinate-related phenomena*. As we see, with an appropriate change of coordinate, the space-time of the black hole displays no special features at its horizon. It is locally more or less flat like everywhere else – except at the real singularity $r = 0$.

Of course, we discover that space-time near the horizon is flat because we use a uniformly accelerated coordinate system to see it clearly – just like on the surface of the Earth inside a free-falling elevator, there is no gravity. Uniform gravity or uniform acceleration don't make the flat space become non-flat. They are essentially equivalent to a change of frame of reference. The reader may want to go back to lecture 1, if necessary, to clearly understand that point. The essential point to remember from the analysis just done is that *nothing special happens at the black hole horizon*.

Why would we be interested in a coordinate system, for a gravitating object, that is mimicking uniform acceleration? Here is why: standing motionless, I am experiencing an acceleration of approximately 10 meters per second per second. It is exactly the same feeling that I would have if I were in free space and the floor under me was being pushed upward with an acceleration of 10 meters per second per second.

Similarly, when you are studying the gravitating black hole from the perspective of somebody standing still, being maintained by a support, let's say, somewhere above the horizon of the black hole, you are effectively doing physics in a uniformly accelerated reference frame in flat space-time.

What does physics feel like there? What does it do there? It does whatever the uniformly accelerated reference frame does. And that is exactly what we experience near the horizon of a black hole: we experience whatever is effectively created by a uniformly accelerated coordinate system – just like you, in Newtonian physics, sitting in your chair. Yet the space-time itself is flat.

Interchange of Space and Time Dimensions at the Horizon

Let's talk about the interchange of space into time and time into space, to which we have already alluded, when we cross the horizon of the black hole. We shall now make full use of the facts we just established – that in the vicinity of the horizon, the gravitational field in the space-time is equivalent to that of a uniformly accelerated reference frame. We defined ξ as

$$\xi = r - 1$$

or, if we put back M and G, ξ would be proportional to $r - 2MG$.

The variable ξ can change sign. How? It changes sign by going from outside the horizon to inside the horizon. It makes sense – at least in the black hole context – to say that ξ can change sign.

This requires some clarifications, though, because in a uniformly accelerated reference frame in space-time, from equations (23), we also have

$$\xi = \frac{\rho^2}{4} \tag{29}$$

How can ρ^2 change sign? Coordinate ρ is a real number and ρ^2 is always positive. But that is not the right way to look at ρ^2.

The right way to look at ρ^2 is to look at the right-hand side of the following equation:

$$X^2 - T^2 = \rho^2 \qquad (30)$$

For a fixed right-hand side, in equation (30), it is the equation of a hyperbola. If ρ^2 is positive, it is a hyperbola the two branches of which are in the first and third quadrants. If ρ^2 is negative, the branches are in the second and fourth quadrants; see figure 13.

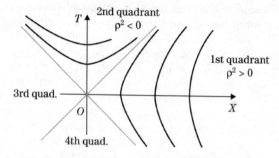

Figure 13: Hyperbolas with positive and negative ρ^2. The point O is not the center of the black hole, but a point on its horizon.

Let's forget about the left and lower quadrants, and concentrate on the right and upper quadrants.

When ρ^2 is negative, this simply corresponds to points in the second, also called upper, quadrant. Going from ξ positive to ξ negative, we are simply passing from the right quadrant to the upper quadrant.

Imagine that we follow the space coordinate ξ till we get to $\xi = 0$. Then, where would we go from there? Let's reformulate the question. We are somewhere on the horizontal axis, therefore above the horizon. The hyperbolic angle ω, to make life simple, is equal to zero. And we are following our way with ξ decreasing toward the origin. At the origin O, which is not the center of the black hole but is the point where we cross the horizon (with θ and ϕ fixed), ξ changes sign. Then, where do we go if we keep moving with ξ now varying in the negative numbers, and ω still zero?

The answer is: we don't go left into the third quadrant; we now go upward into the second quadrant.

On the X-axis, in figure 13, $\omega = 0$ and $\xi > 0$. And on the T-axis, $\omega = 0$ and $\xi < 0$.

What happens is that a space-like coordinate, namely ξ, jumps to become a time-like coordinate. Indeed, when ξ is positive, a variation in ξ entails an interval displacement of the space-like type. Then, when ξ is negative, a variation in ξ entails an interval displacement of the time-like type.

Let's now turn to the coordinate ω. In the first quadrant, the hyperbolic angle ω is a time-like coordinate. When it varies alone, on a given hyperbola, we move on a time-like interval (with a slope $> 45°$). What happens to it after we pass the origin O? Now, in the second quadrant, letting ω vary while ξ is fixed, we move along a hyperbola of the second quadrant that is along a space-like interval, as seen in figure 13. In summary, when we pass O, that is, when we cross the horizon, the space and time dimensions in the coordinate system (ξ, ω) are interchanged.

Let's not be mistaken: ξ remains ξ, and ω remains ω. But the former, which was space-like, becomes time-like. And the latter, which was time-like, becomes space-like.

These considerations concern *coordinates* of events in space-time. Nothing physically meaningful happens to particles following trajectories in space-time – except when they hit the singularity at the center of the black hole, but that is another story. In the next section we shall study several trajectories crossing the horizon. We shall see in more detail what makes the singularity singular.

Black Hole Singularity

Instead of thinking about a particle crossing the horizon, we now think about the black hole itself.

The coordinate ξ is simply $r - 1$. Decreasing r until we got to $r = 1$ and then $r < 1$, while keeping t fixed, consisted in reaching and then crossing the event horizon. This also consisted in reaching and crossing the origin in figure 13. Then continuing to let r decrease toward zero, we are now moving up along the vertical axis. That is the nature of these coordinates, and is not the most important point from a physical point of view.

Eventually we will reach $r = 0$, and that is what we want to study now. So let's redraw the previous figure with less information; see figure 14. And let's now focus on the singularity.

At $r = 0$ something nasty happens. Where is $r = 0$? It is in the upper quadrant on the hyperbola shown in figure 14. We are then behind the horizon, inside the sphere of radius $r = 2MG$. The coefficient $(1 - 2MG/r)$ has changed sign and gone to $-\infty$.

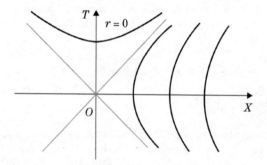

Figure 14: Singularity $r = 0$. It is not a vertical straight line, but a time-like curve in the upper quadrant.

Lo and behold: the singularity at $r = 0$ is not a place, it is a time! More accurately, it is not a single place but many places all on a time-like curve. The surface $r = 0$ is not what we would normally think of. Normally a place corresponds, in the Minkowski diagram, to a vertical line. But this is not true of the singularity of a black hole at $r = 0$. In this case it is the hyperbola in the upper quadrant in figure 14. And it is more time-like than space-like.

This is difficult to figure out and of course is at first confusing. But this is what the black hole really is. You will get used to it, if you think about it and practice with it.

An important point to note about the interior of the black hole is this: once we got past the horizon, we cannot avoid eventually running into the singularity. Why? Because we cannot avoid the future.

Indeed, there is no way to escape from the future the way we can escape from a place we dislike. Think of a place in front of you that you don't want to run into: you can avoid it by just going around it. Once we are past the origin in figure 14, and we are inside the horizon, which is in the upper quadrant, no matter what we do we will run into the hyperbola $r = 0$.

The singularity is not avoidable the way it would be in Newtonian physics. Indeed, in Newtonian physics the center of the coordinates is a place. We can go around a place, but we cannot go around a time. That is the nature of the black hole singularity.

What about the horizon? The horizon is the point O in figure 14. But we are going to think about it a little differently eventually. We are going to think about *the whole straight line at 45°* as being the horizon.

Figure 14 illustrates clearly the idea that it takes an infinite amount of time for an object to fall through the horizon – at least an infinite amount of *coordinate time T*. But still, it falls through in its own time keeping, which is its proper time. And with that time it falls through in a finite amount of time. Let's redraw the figure and see if we can understand that; see figure 15. And now rather than objects we will study people falling into the black hole.

Consider the two famous protagonists in physics: Alice and Bob. Alice, one way or another, manages to stay outside the horizon of the black hole at some fixed distance ρ from the horizon. And she pushes Bob or at least observes him fall into the black hole.

In figure 15, the origin O is the point on the horizon below her, as in figure 14. The straight lines from the origin are time slices of constant ω. Remember that ω is like time. The relationships between standard Minkowski coordinates and hyperbolic coordinates are given by the following familiar equations:

$$X = \rho \cosh \omega$$
$$T = \rho \sinh \omega \qquad (31)$$

Alice's trajectory is the hyperbola in figure 15. Even though she doesn't move, it is not a vertical straight line because of the Schwarzschild metric, which corresponds to a gravitational field. We saw that, at any point except the center of the black hole, it is everywhere locally equivalent to a uniformly accelerated frame, even in the vicinity of the horizon. Therefore a motionless point above the horizon follows a hyperbola like we studied in lecture 4.

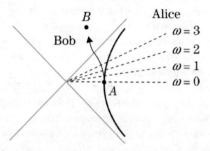

Figure 15: Alice stays outside the black hole horizon at a fixed position. Her trajectory in space-time is the hyperbola. On the other hand, the unfortunate Bob falls in.

Alice's ρ doesn't change, and the time, measured by ω, ticks as she passes each radial dotted line in the picture. Remember that ω is proportional to her proper time since $d\rho = 0$; see equation (22). The variable ω asymptotically goes to infinity when the radial lines become closer and closer to the 45° line. *For Alice* that line (or cone in several spatial dimensions) with 45° slope corresponds to infinite time. Notice that the coordinate time T, i.e., the time of the physicist observer of the whole situation if you like, becomes infinite if and only if ω becomes infinite; see equation (31).

If we consider now Bob's unfortunate trajectory, it is not until an infinite amount of Alice's proper time has elapsed that Bob has

gone through the horizon. In other words Alice never sees Bob
go through the horizon. According to her reckoning, it takes an
infinite amount of ω before Bob falls into the black hole. That, as
just said, also corresponds to an infinite amount of T.

On the other hand, just looking at the diagram in figure 15, we
can see immediately that the amount of proper time it takes Bob
to go from point A to point B is finite. Bob goes through the
horizon in a finite time. The surface $X = T$ *is* the horizon. We
will come back to it.

It is only due to the coordinates she uses that Alice is led to
reckon that Bob takes an infinite amount of time to fall through
the horizon. Bob says: "No, I watched it on my clock, it only
takes a finite amount of time." The reason is the difference between proper times and coordinate time. Alice uses her proper
time, which is linked to coordinate time by $T = \rho \sinh \omega$, while
Bob uses his own proper time.

We can figure out how Alice "sees" Bob. It requires somehow
for Alice to look back in time, because when we speak of looking,
we refer to light arriving to our eyes, and light takes time to travel.

But let's talk of "looking" in a different way: when we say Alice
looks at Bob at time ω, we mean she figures out what does Bob
do at this simultaneous time ω, in other words, where is Bob on
the radial line of value ω. Well, we know from figure 15 where he
is. We deduce that Alice sees Bob slow down. Each heartbeat of
Bob takes longer in Alice's time.

Now let's turn to the question of how Bob sees Alice. We apply
the same idea: by what Bob "sees" of Alice, we mean what happens to Alice at the same Bob's time, when he "looks back," that
is, going into the past coordinate time, but simultaneous for him.

Those lines of simultaneous times for Bob are $-45°$ lines that link
points on Bob's trajectory and Alice's trajectory, as in figure 16.
We see that Bob doesn't see anything special. He sees Alice before
he plunges through the horizon. He sees Alice while he is at the
horizon. And he keeps on seeing her afterward without problem.

While on his trajectory, on which he does not notice anything particular happening when crossing $r = 2MG$, Bob doesn't see Alice shrink. True, he does see Alice accelerate away from him. But that's all. At any point, he sees Alice perfectly normally, except that she is accelerating away from him.

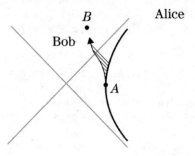

Figure 16: How Bob sees Alice. The light rays reaching Bob, coming from Alice, come from the past, therefore they are at a 45° angle turned to the left.

Notice that when Bob is at B, he can see Alice. Whereas Alice can never see Bob at point B. In other words, it is a very asymmetric situation.

What happens when Bob hits the singularity? By that point, the tidal forces will become so large that Bob will no longer be with us. Bob will have been destroyed by the time he gets to the center of the black hole field. That is why we don't think about trying to answer the question: what does Bob see when he gets to, say, a negative r? By the time he gets to negative r, he has experienced infinitely strong tidal forces and there is no more Bob.

Notice that in the upper quadrant Bob will probably use r for time or something like r. Inside the horizon, where r varies from $r = 2MG$ to $r = 0$, Bob's clock would have more to do with r than with t of equation (2). And t would be more like position. But, as I said, Bob doesn't feel any funny thing happening with space or time. His clock doesn't become a meter stick, and his meter stick doesn't becomes a clock. Nothing of the sort happens.

He just sails through.

Yet the line at 45° in figure 16 and the previous ones is rather special. Let's see what can and cannot happen once something – a person, an object, a particle, even a photon – has crossed the horizon.

No Escaping from a Black Hole

We repeated many times and showed that "nothing special happens at the horizon." Yet, to dispel the idea that the horizon is like any other place, let's now explain in detail why it is not quite so. Indeed, anything that crossed the horizon can no longer escape. It is somehow trapped into the black hole. And it is doomed.

Remember, in the coordinates that we are using, light moves with a 45° angle. Therefore light cannot escape from the upper quadrant in figure 17.

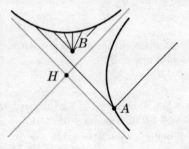

Figure 17: Trajectories starting inside or outside the horizon. (We changed the name of point O to point H.)

All it can do is eventually hit the singularity. And anything that is moving slower than the speed of light has a slope closer to the vertical, and will also hit the singularity. Consequently, anybody who at some point is in the upper quadrant is doomed.

On the other hand, somebody in the right quadrant has the possibility of escaping.

Let's focus again on a light ray because it is simple. Consider one originating in the right quadrant, i.e., outside the black hole. If it is moving radially outward, it will escape. It will simply just keep going. If it is pointing radially inward, it is of course also doomed. The light ray will hit the singularity, as in figure 17.

There is also everything in between. If a light ray is pointing out of the page, depending on whether it is inside or beyond the photon radius, it will fall in or not.

Figure 17, and its variants figures 15 and 16, are pictures you should familiarize yourself with, until they no longer have any secrets. If you want to understand and be able to resolve weird paradoxes about who sees what in the black hole, I recommend that you always go back to these diagrams.

A good understanding of black holes is a prerequisite to understanding general relativity. The reason is that black holes are the simplest, ideal form of massive bodies. Their mass is concentrated at a point. It is the equivalent in relativity of point masses in Newtonian physics.

Point masses in Newtonian physics give rise to Newtonian gravitation, trajectories solved with Newton's equation, etc. Their physics has a singularity at the central point itself, but nowhere else. In relativity the analog of point masses are black holes, and their metric. It is the Schwarzschild metric. Black holes display of course a singularity at their center, but also strange phenomena at $r = 2MG$ and below. We stressed that the peculiar phenomena happening at $r = 2MG$ only come from the coordinates, not from real physical strangeness. A black hole horizon, nonetheless, is special in the sense that it is the limit of no return.

The next two lectures will be devoted to strengthening our understanding of these phenomena and more generally deepening our knowledge of black holes, which are an indispensable preliminary topic to understand general relativity.

Lecture 7: Falling into a Black Hole

Lenny: *Today we shall study black holes in even more depth. Ready for the trip?*

Andy: *You said that there is no escape from the future ... so I guess I have no choice. The interior of a black hole reminds me of the time when a friend of mine took me on the Ferris wheel at Santa Cruz. But she went with me and we both came back.*

Introduction

We shall review, become familiarized with, and deepen what we began to learn in the previous lectures, namely what happens around a black hole, outside the horizon, at the horizon, inside the black hole, and at the singularity. So there will be some repetition of what we have already said, but it should serve a useful purpose.

Schwarzschild Metric, Event Horizon, and Singularity

Let's get back to the Schwarzschild metric, which describes the geometry of space-time caused by the presence of a black hole.

To begin, we shall simplify the metric by rescaling the coordinates. This won't change anything about the physics of course, but it will make the mathematics simpler. Let's first write it as we learned it in the previous lectures, but with one slight change. We put primes on the time and radial coordinates, so that after the change of variables, the new variables will be as usual $(t, \ r, \ \theta, \ \phi)$:

$$d\tau^2 = \left(1 - \frac{2MG}{r'}\right)dt'^2 - \left(\frac{1}{1 - \frac{2MG}{r'}}\right)dr'^2 - r'^2 d\Omega^2 \quad (1)$$

This is exactly the equation we learned earlier, except for the slight change of notation.

Let's give a new name to the Schwarzschild radius $2MG$:

$$R_S = 2MG \quad (2)$$

The index S of R_S stands for "Schwarzschild." Next, we rescale the time and radial coordinates. We *define* r and t as follows:

$$r = \frac{r'}{R_S}$$
$$t = \frac{t'}{R_S} \quad (3)$$

Normally radii have units of length. Moreover if we set $c = 1$, and here we use the convention that it implicitly multiplies the time variable t', time also has units of length.[1] But when we divide a length by a length, we get something dimensionless. So our new r and t, defined by equations (3), are dimensionless.

The equation of the metric takes the simpler universal form

---

[1] Sometimes one meets the implicit c as a dividing factor in front of the spatial variables, in which case the time variable retains units of time. But here we want the time variable t' to be a length. Then t, defined by equation (3), will become dimensionless.

$$d\tau^2 = R_S^2 \left[\left(1 - \frac{1}{r}\right) dt^2 - \left(\frac{1}{1 - \frac{1}{r}}\right) dr^2 - r^2 d\Omega^2 \right] \quad (4)$$

Inside the brackets is the metric that would have been written down in the first place had $2MG$ been equal to 1.

What we find out from this exercise is that basically the metrics of all Schwarzschild black holes are the same, except for an overall factor proportional to the square of the radius of the black hole. Everything inside the brackets is dimensionless. And $d\tau^2$ has the same units as R_S^2.

For most purposes in studying the geometry of a black hole, we can simply set R_S equal to one. We are in the same situation as when we study the metric of a sphere. There are big spheres and small spheres but, up to a factor equal to their radius, they all have the same geometry.

Now let's turn to an important question: what is the curvature of space-time near a black hole?

The curvature tensor, which we studied in lecture 3, is the tool that tells us, at any point, whether the space-time is flat or curved. Curvature is analogous to tidal forces; it is the tendency for the geometry to distort objects. Tidal forces are the effect of curvature. You may say that, in some sense, they *are* curvature.

The expression for the curvature tensor is

$$\mathcal{R}_{srn}^{\;\;\;t} = \partial_r \Gamma_{sn}^t - \partial_s \Gamma_{rn}^t + \Gamma_{sn}^p \Gamma_{pr}^t - \Gamma_{rn}^p \Gamma_{ps}^t \quad (5)$$

Our question is: how big are the components of the curvature tensor, let's say at the horizon? (The horizon corresponds to $r = 1$.) At first sight one might think they are infinite, but that's not so.

As a preliminary question let's ask ourselves: what are the units of curvature? Remember that a Christoffel symbol is expressed, in terms of the metric components, with the formula

$$\Gamma_{mn}^t = \frac{1}{2}\, g^{rt} \left[\, \partial_n g_{rm} + \partial_m g_{rn} - \partial_r g_{mn} \,\right] \quad (6)$$

Also remember that

$$dS^2 = g_{\mu\nu}dX^\mu dX^\nu \tag{7}$$

From equation (7), it follows that $g_{\mu\nu}$ is dimensionless, and so is its inverse $g^{\mu\nu}$. Then, since the Christoffel symbols, in equation (6), contain spatial derivatives of g, they have units inverse of length. We conclude, from equation (5), that the curvature tensor components have units inverse of length squared. Using the standard notation with brackets for dimension, this is expressed by writing

$$[\mathcal{R}] = \frac{1}{[\text{length}]^2} \tag{8}$$

Let's come back to our question: what is the curvature at $r = 1$? We are going to use a nice argument resting only on dimensions to find the answer, i.e., we shall carry out what is called a dimensional analysis. Looking again at the metric, the only quantity on the right-hand side of equation (4) that has units is the radius R_S, which is a length. Everything else is dimensionless. Therefore when we calculate the curvature, the only possibility for the curvature to have units inverse of length squared is if it is inversely proportional to R_S^2 itself, that is, to the Schwarzschild radius squared.[2] We write it

$$\mathcal{R}_{\text{Horizon}} \sim \frac{1}{R_S^2} \tag{9}$$

where $\sim$ stands for "proportional to," with a fixed proportionality ratio, the same for all black holes.

For different black holes, the bigger the black hole, the smaller the curvature at its horizon, in other words the flatter it is there. So the tidal forces at the horizon of a large black hole are less severe than the tidal forces at the horizon of a small black hole. A small black hole is a much nastier thing near its horizon than a very big black hole.

---

[2]This should not be a surprising fact. Recall that the Gaussian curvature of a sphere of radius r is everwhere $1/r^2$.

Now let's concentrate on the geometry whose metric is represented by equation (4). To begin, let's throw away the R_S^2, that is, let's replace it by 1. We consider, if you like, a "unit black hole," a black hole of radius 1.

We will introduce a new coordinate to replace r. Figure 1 shows the r-axis.

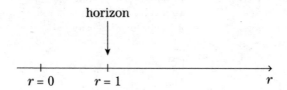

Figure 1: Radial coordinate r.

We want to construct a new coordinate that measures *proper distance from the horizon*. How do we calculate the distance from the horizon? We go back to the expression of proper distance[3]

$$dS^2 = -\left(1 - \frac{1}{r}\right) dt^2 + \left(\frac{1}{1 - \frac{1}{r}}\right) dr^2 + r^2 d\Omega^2 \tag{10}$$

Now let's move away from the horizon radially outward. Consider two points A and B at distance respectively r and $r + dr$ from the point of abscissa $r = 0$, as in figure 2. What is the proper distance between these two points?

To find out, we use equation (10).

First, notice that we compare two points at the same time, so $dt = 0$. Secondly, we move radially, so Ω doesn't change either. Therefore between A and B, the proper distance squared is simply

$$dS^2 = \frac{dr^2}{1 - \frac{1}{r}}$$

---

[3]Recall that when we work with $c = 1$, the square of the differential of the proper distance, dS^2, is simply the opposite of $d\tau^2$ given by equation (4).

which can be rewritten as

$$dS = \sqrt{\frac{r}{r-1}}\, dr \qquad (11)$$

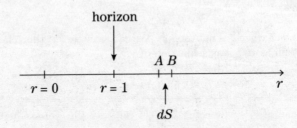

Figure 2: Proper distance dS between A and B.

Let's call the proper distance from the horizon ρ, as in figure 3.

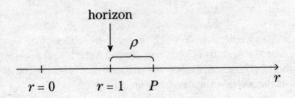

Figure 3: Proper distance ρ from the horizon.

How do we find the proper distance between the point P and the horizon? We integrate equation (11):

$$\rho = \int_{u=1}^{u=r} \sqrt{\frac{u}{u-1}}\, du \qquad (12)$$

The integral is not hard to do, but in what follows we will only be interested in the region near the horizon where $u \approx 1$. In that region the integral can be approximated by

$$\int_{u=1}^{u=r} \frac{du}{\sqrt{u-1}} = 2\sqrt{u-1} \ \Big|_{u=1}^{u=r} = 2\sqrt{r-1} \qquad (13)$$

Thus we found the proper distance ρ from P to the horizon as a function of r:

$$\rho = 2\sqrt{r-1}$$

This holds when r is not too big. We can easily invert this relation to obtain as well r as a function of ρ:

$$\frac{\rho^2}{4} + 1 = r \qquad (14)$$

The metric given by equation (11) can then be expressed either in terms of r or in terms of ρ.

To give equations a nice form, we also redefine the time variable. We define ω as

$$\omega = \frac{t}{2} \qquad (15)$$

Now let's re-express dS^2 of equation (10) with the variables ρ and ω instead of r and t. The second term, which is $r/(r-1) \ dr^2$, is simply $d\rho^2$. We see this from equation (11), or its squared variant, remembering that we decided to call the proper distance ρ.

Regarding the first term $-(r-1)/r \ dt^2$, it can be expressed as a function of ρ times $d\omega^2$. For reasons that will become clear in a moment, let's write this function as $-F(\rho)\rho^2$.

With the change of variables from $(t, \ r)$ to $(\omega, \ \rho)$, we can re-express the metric. Let's switch back to $d\tau^2$, which is just $-dS^2$ and which we have generally chosen to express the metric; see equation (1). We get

$$d\tau^2 = F(\rho)\rho^2 d\omega^2 - d\rho^2 - r(\rho)^2 d\Omega^2 \qquad (16)$$

It is possible to calculate explicitly the last coefficient $r(\rho)^2$, but we don't need to. The form of the metric is sufficiently interesting as it is in equation (16) if we also have some knowledge about what $F(\rho)$ and $r(\rho)$ are like.

So far we just went over matter that we already cursorily saw in the last section of lecture 6, entitled "Hyperbolic coordinates revisited." But we are going into more detail. Moreover we now use explicitly the proper distance, which we denoted ρ, whereas in lecture 6 we worked differently: we stayed near the horizon, at $r \approx 1$, and worked directly with the quantity $r - 1$, which we denoted ξ; see equations (26) to (28) of lecture 6.

Where is the horizon in terms of ρ? It is at $\rho = 0$, because ρ is the (proper) distance from the horizon.

Poring over equation (16), we can say a number of things. First of all, when ρ becomes very big, far away, it is easy to show that

$$\lim_{\rho \to +\infty} F(\rho)\rho^2 = 4 \qquad (17)$$

That just reflects the metric far from the black hole. It is a first thing. Next, let's look at what happens when ρ goes to zero. That is moving leftward in figure 3 right up against the horizon. Again we easily establish that

$$\lim_{\rho \to 0} F(\rho) = 1 \qquad (18)$$

The first term in the expression of $d\tau^2$, in equation (16), becomes $\rho^2 d\omega^2$. You may find that familiar. That was the reason, by the way, for defining the term function of ρ in front of $d\omega^2$ as $F(\rho)\rho^2$. Indeed, that way, close to the horizon $F(\rho)\rho^2$ is just ρ^2, or equivalently $F(\rho)$ is just equal to one.

Finally, let's look at the limit of $r(\rho)$ when ρ goes to zero. That we can work out. As we said, $\rho = 0$ means we are at the horizon. There $r = 1$, but we can do better. We can show that

$$\lim_{\rho \to 0} r(\rho) = 1 + \frac{\rho^2}{4} \qquad (19)$$

This nonstandard way to write a limit means, if you prefer, that $r(\rho)/(1 + \rho^2/4)$ tends to one.

Equations (17), (18), and (19) are basically everything that we need to know to study the metric.

Now the most important thing is to look at equation (16) when ρ is small. It is approximately

$$d\tau^2 = \rho^2 d\omega^2 - d\rho^2 - d\Omega^2 \qquad (20)$$

That is exactly the same as the metric of flat space but expressed in hyperbolic polar coordinates.

Let's review quickly hyperbolic coordinates in the vicinity of a point H on the horizon; see figure 4. Recall that there are two important quadrants in the picture (we are talking about the four quadrants defined by the *diagonals*, not by the X- and T-axes): the right quadrant of events located farther than the horizon, and the upper quadrant of events located inside the horizon.

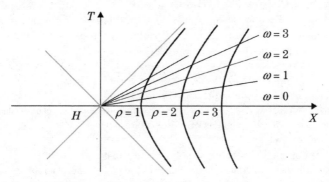

Figure 4: Hyperbolic coordinates $(\omega,\ \rho)$ near the horizon H. The pair of variables $(X,\ T)$ are the Minkowski coordinates used by an observer near H who is free-falling in the gravitational field of the black hole. He perceives space-time as flat, and a fixed object as accelerating.

Look first at the right quadrant. The first hyperbola corresponding to $\rho = 1$ is the trajectory over time of a particle fixed at one unit of ρ away from the horizon. The second hyperbola is the trajectory of a particle fixed at $\rho = 2$, etc. That is just looking at the metric of equation (16) and noticing that, apart from the coefficient $F(\rho)$, near the horizon it just has the form of flat space in hyperbolic coordinates that we studied in lecture 4.

The lines of constant ω are the straight lines fanning out of the origin in figure 4, that is, the point, or event, H on the horizon.

We drew $\omega = 0$, which is the horizontal axis, and $\omega = 1$, $\omega = 2$, $\omega = 3$. Where is $\omega = +\infty$? It is right along the light cone, which on the diagram is the line at 45°. Remember what is ω: it is simply $t/2$. We had defined t and r as follows:

$$X = r \cosh t$$

$$T = r \sinh t$$

Thus we constructed – or, you may say, discovered – some coordinates, (ω, ρ), in which time is like a hyperbolic angle. Time-infinity[4] is the light cone whose generatrix is at 45°.

What about the upper quadrant? We already talked about it in the last lecture. It is the region where $r < 1$. In that region the sign of the first two terms in equation (10), or equivalently in equation (16), interchange. In the upper quadrant ρ^2 is negative. We can see this from equation (19).

In the upper quadrant the curves of constant ρ, or equivalently constant r, are the hyperbolas shown in figure 5 in the next section. We drew $r = 1/2$ and $r = 0$. When we have several spatial variables, the hyperbolas are hyperboloids.

Nothing crazy physically is going on in the upper quadrant. Time has not become space, space has not become time. We have just introduced *coordinates* that have the funny property that when you go from the right quadrant to the upper quadrant, what was time-like becomes space-like and conversely. It is entirely a coordinate artifact.

Despite this coordinate artifact, there are many interesting points about the diagram. We will describe them in this lecture and the next one. To start with, let's mention that if we go far away from H, we have to remember that $F(\rho)$ in equation (16) is not just 1; and $r(\rho)$ is not just 1 either. That means the metric along the ω direction, $F(\rho)\rho^2$, does differ from flat space. But the difference shows up in the way $F(\rho)$ varies as we move away from the black hole. The same comment applies to $r(\rho)$ in front of $d\Omega^2$.

---

[4]By *time-infinity*, we mean the line fanning out of H corresponding to infinite time ω; see figure 4.

Fundamental Diagram of Space-Time near a Black Hole

Figure 5 presenting space-time viewed by a free-falling observer near the horizon of a black hole is the *fundamental diagram of space-time near a black hole*. We must understand it very well.

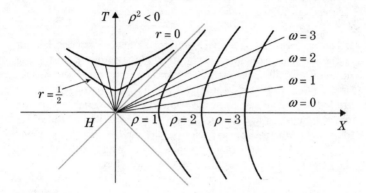

Figure 5: Constant ρ's and constant ω's in the right quadrant and in the upper quadrant; H is a point on the horizon of the black hole.

The origin H corresponds to $\rho = 0$. It is also $r = 1$, i.e., it is not the center of the black hole. What happens as we move up into the upper quadrant and r decreases? Eventually we hit the nasty point where $r = 0$. In our diagram it is an entire branch of a hyperbola. But that is an abstract representation, just like a fixed point in space in the classical Minkowski diagram is represented by a vertical line. However, we are no longer just in the presence of a coordinate glitch, we are at a real geometrical and physical singularity.

If we calculated the curvature near the hyperbola $r = 0$ (or the hyperboloid in several spatial dimensions), it is not too hard to figure out what we would find. As r goes to zero, the sphere is getting smaller and smaller and the curvature larger and larger. It becomes infinite on the hyperbola $r = 0$. This is a true singularity. It is a place of infinite tidal forces. Basically it is a place where we don't want to be.

The problem with the singularity is that once we are in the upper quadrant, i.e., once we have gone through the horizon and are "inside the black hole," we can't avoid eventually hitting it. Indeed, it is not really a place. As already discussed last lecture, it is really, in a sense, a time. We can avoid an obstacle in space – we go around it – but we cannot avoid "hitting the future." We can escape from things, even from freedom, but not from the future.

If you are orbiting in a space station far away from the horizon of the black hole, as long as you don't do anything doltish – like jump off and allow yourself to get sucked in by gravity – you are safe in your capsule. Nothing bad will happen to you. But if you are foolish enough to think "I want to explore what is in there" and decide to go and see, as soon as you pass the horizon of the black hole, you are doomed. Not only is there no way you can get out, but there is even no way you can avoid eventually flying into the singularity and being destroyed by tidal forces.

In order to get out you would have to exceed the speed of light. That corresponds to the straight lines with a 45° slant in figure 5, or to the corresponding light cones if we add the other spatial dimensions. Once you are in the upper quadrant, the best you can do to try to escape – if we assume that nothing can exceed the speed of light – is to follow a 45° trajectory. Yet they all run into the hyperbola $r = 0$ in a finite amount of your proper time.

Where is the point of no return? It is actually the entire diagonal line at 45°. To understand why, let's ask ourselves: what is that light cone? In the right quadrant each hyperbola is at a different value of ρ. The entire hyperbola corresponds to its value of ρ. In figure 5 we have drawn the hyperbolas $\rho = 3$, $\rho = 2$, $\rho = 1$. At point H, $\rho = 0$. But in fact ρ is equal to zero all along the limit hyperbola that consists of two lines at 45° and −45°.

There is something very different about this kind of geometry than ordinary geometry. Suppose we drew ordinary geometry, as in figure 6, and asked: where is the point $r = 0$? Well we know that there is only one point where $r = 0$. It is the origin. Where is the point where $r = \epsilon$? It is not a point, it is a tiny circle around O. As ϵ goes to zero, the circle eventually becomes just point O.

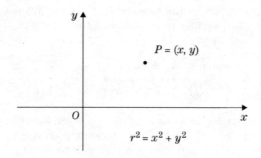

Figure 6: Flat plane in ordinary Euclidean geometry. The quantity r is the distance from O to P.

It is quite different in the geometry of space-time. Where is $\rho = \epsilon$? It is an entire hyperbola very close to the light cone. As ρ gets smaller and smaller, it tends to the light cone, not just to the point H. And $\rho = 0$ is the horizon.

Anyone who happens to be beyond a point on the diagonal line, meaning by that in the upper quadrant, is "inside the black hole," in the sense of inside its horizon. As we can see just looking at figure 5, such a person is doomed.

To summarize, the point of no return is indeed $r = 1$, or equivalently $\rho = 0$. But it is not only the point H. It is the whole line at $45°$. This whole line in figure 5 is the horizon of the black hole.

Notes on the Fundamental Diagram

As already said, it is very important to grasp perfectly the fundamental diagram of space-time near a black hole, shown in figure 5, in order to understand gravitational fields created by massive bodies, and eventually the theory of general relativity. Here is a list of notes to strengthen our understanding:

1. In the upper quadrant, the hyperbolas of constant r, or equivalently constant ρ, correspond to $\rho^2 < 0$. Therefore the ρ's are imaginary.

2. In the same upper quadrant, the straight lines coming out of H are still lines of constant t, or equivalently constant ω. But they are now space-like. And the hyperbola of a given r is time-like.

3. The coordinates (X, T) are those naturally used by an observer who is near the horizon of the black hole, and is free-falling in the gravitational field of the black hole.

4. Indeed, even though space-time *is not flat*, the observer free-falling through the horizon perceives it as flat. Therefore they might find it convenient to use the rectangular coordinates (X, T) to chart what is going on around them. T is by definition the proper time of the observer. It is sometimes called the *coordinate time*. Coordinate X of course measures distances with the stick of the observer.

5. A particle that, unlike the unfortunate observer, is at a fixed distance from the black hole, for instance at $\rho = 2$ from its horizon, follows a trajectory in space-time that, viewed in the frame of reference of the observer, is the hyperbola in the right quadrant indexed with $\rho = 2$.

6. Time T and time t are different. Time T is the time of the observer. Time t is the time of any particle fixed relative to the black hole. Such a particle is living on one of the hyperbolas in figure 5. More accurately, for each particle, its proper time is proportional to t, or equivalently proportional to ω, which is $t/2$.

7. It is important to understand that, just as in Newtonian physics we need a reference frame to represent space and time – time being a neatly separate and universal coordinate – in relativity we also need a reference frame to represent space-time. Since, due to massive bodies in the universe, there are no genuine uniform fields, no frame is clearly superior to any other. However the frame of reference of a free-falling observer remains particularly convenient.

The coordinates (ω, ρ) have a name: they are called the *Kruskal coordinates*, named after the American mathematician and physicist Martin Kruskal (1925–2006).

History of Black Holes

When, in late 1915, Karl Schwarzschild wrote down the metric now bearing his name, he didn't know that the horizon was a horizon. For some time, as far as he – as well as Einstein and all the other people who studied this subject – could tell, $r = 1$ was a nasty place where two coefficients in the metric given by equation (4) changed sign and also one of them – the one in front of dr^2 – went through infinity. They said: "Oh dear, this coefficient is divergent at $r = 1$, something bad is happening there." The conclusion was that the horizon was some sort of singularity. In other words, the first people who studied black holes did not realize that the horizon was a smooth nonsingular ordinary place.

Sometime in the 1950s, David Finkelstein[5] realized that the horizon of a black hole was the *point of no return*, but also that at the horizon itself nothing nasty happened yet to someone falling through. Only some finite proper time later would that person be annihilated at the singularity. He rediscovered coordinates that had already been written down by Arthur Eddington.[6] So they were called the Eddington–Finkelstein coordinates. They are not exactly the same as the Kruskal coordinates, but are similar.

Martin Kruskal was not a specialist of relativity. He was a plasma physicist but was very good at equations and very good at changing coordinates. He loved to change coordinates. Somebody showed him the Schwarzschild metric. He tried many changes of coordinates and eventually, in 1960, found these coordinates (ω, ρ) in which the metric has the nice form of equation (16), that is,

$$d\tau^2 = F(\rho)\rho^2 d\omega^2 - d\rho^2 - r(\rho)^2 d\Omega^2$$

He proposed the now familiar diagram that accompany them, that is figure 5, which we called the fundamental diagram of space-time near a black hole.

---

[5]David Finkelstein (1929–2016), American physicist.

[6]Arthur Eddington (1882–1944), English astronomer. He headed the expedition that, during the solar eclipse of May 29, 1919, photographing the deflection of light rays by the mass of the Sun, first confirmed Einstein's theory. Since then the predictions of the theory have been verified many times with more accuracy. Nowadays general relativity plays an important role in some applications like, for instance, the Global Positioning System.

Before being called a black hole, this kind of object was called a collapsed star or massive collapsed star. The term *black hole* was coined by John Wheeler.[7]

John Wheeler was a very nice and very sweet man. He was a good friend of mine too. He was politically very conservative, contrary to me. His political conservativeness had to do with one thing and only one thing: he was concerned about the Soviet Union having any nuclear weapons. So he was very anti–Soviet Union and especially anti-Soviet expansionism. We would argue about it, not so much me, but some of my friends. He was, however, a very thoughtful, gentle person.

His political conservativeness did not extend to social issues. I remember once we were sitting in a café in Valparaiso, Chile, John – he was like 85 years old at the time – my wife, and I. While sitting, he started to look restless. I said: "What's the matter John, do you feel well?" He said: "I feel fine, I just want to get up and take a walk." I said: "Where are you walking? Do you want me to take a walk with you?" He said: "No, no, I'm gonna take a walk by myself." I asked: "What are you going to do John?" He answered: "I want to check out the bikinis." So he was not a social conservative.

In his first paper on the Schwarzschild metric, he coined the term *black hole*. That caused a stir. The *Physical Review* did not want to publish it. It was before my time as an active scientist, or just about that time, but I knew there had been a problem. I didn't know what it was. I learned it was not just the *Physical Review* being its usual conservative self. To be true, they were being hyper-conservative: they thought that the term *black hole* was obscene. So they refused to publish the paper at first. And John fought and fought and fought with them, and eventually won. Then just to get back at them in his next paper he wrote that "black holes have no hair."

Incidentally, what does it mean for black holes not to have any hair? It means that if you take a non-rotating black hole, gravity is so strong that it will always pull it together into a perfect sphere.

---

[7] John Wheeler (1911–2008), American theoretical physicist.

Even if it starts very asymmetric, like two rocks coming together, after a very small amount of time, the horizon will pull itself into a sphere and become indistinguishable from a perfect one. John called that characteristic of not having any visible structure nor any visible defects on it, the property of having no hair.

If the black hole is rotating, however, the sphere can get deformed. It can become an oblate spheroid. But the nature of the oblate spheroid depends only on the angular momentum.

Now, let's talk about things or people falling into a black hole, and in particular what communications are possible and what are impossible.

Falling into a Black Hole

Back to the diagram in figure 5, you might ask what do the left and bottom quadrants represent. We are going to see that this other half of the diagram, the half on the left and below the $-45°$ line, has no real significance. But for the moment we are mainly interested in the upper and right parts. The exterior of the black hole is the right quadrant. The interior of the black hole is the upper quadrant.

Let's redraw the picture; see figure 7. It is Alice's turn to fall in. For simplicity of notations, we revert to coordinates (t, r), which are equivalent to (ω, ρ), the close correspondence being given by equations (14) and (15). Bob stands somewhere outside the black hole at a fixed position. Therefore his trajectory is a hyperbola of constant r as his proper time ticks proportionally to t.

The horizon is $t = +\infty$. It is very strange but this line, which we thought of as being a place, also has the character of being a time. If you prefer to use the time-like variable ω, it is where $\omega = +\infty$.

Looking at the diagram, we can see that while Alice is falling in, she doesn't pass the horizon until t equals infinity. It is in that sense that an in-falling object never passes the horizon. *But it is a statement with the time of the outside observer Bob.* Indeed, Bob cannot see an object nor a person fall through the horizon.

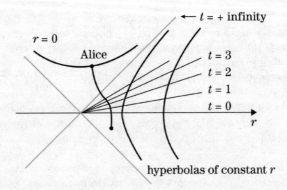

Figure 7: Alice is falling into the black hole. For simplicity we now use the coordinates t and r. Remember that t is twice ω, see equation (15), and r is some function of ρ, see equation (14).

Nevertheless Alice of course does pass the horizon.[8] We can see her in the diagram sailing into the singularity. Nothing special happens to her as she passes the horizon. It is just the strange set of coordinates that says that $t = +\infty$ when Alice crosses the horizon of the black hole.

So you might think: well, the fact that Alice never gets in doesn't really mean anything to anybody. Yet, on the other hand, Bob is out there, staying at a fixed position, that is, on a hyperbola of constant r. And Bob is watching Alice. Let's examine more precisely what he sees of Alice, what it means "to see Alice." How does he see Alice?

"To see Alice" at time t means for Bob to receive a light ray emitted by Alice sometime earlier[9] and arriving at Bob at time t. In other words, Bob "looks into the past"; see figure 8. For instance, at time t_1, when Bob is at point P_B, he sees Alice at point P_A. At time t_2, when Bob is at point Q_B, he sees Alice at point Q_A. Remember that light travels at a 45° angle in our representations.

---

[8]Notice the ambiguity of this statement. It is inherent to the theory of general relativity. For Bob she never crosses the horizon. For herself she unfortunately does.

[9]The lines of simultaneous time for Bob are the straight lines fanning out from the origin of the diagram.

Figure 8 makes it clear that Bob will never see Alice cross the horizon, that is, the line $t = +\infty$. As long as Bob stays outside the black hole and looks back, what he will see is Alice getting closer and closer to the horizon, but never passing it. So in fact, you may say that it is not just a coordinate artifact that Alice "never crosses the horizon." What Bob can see or not is a physical fact. As far as he is concerned, Alice never passes the horizon.

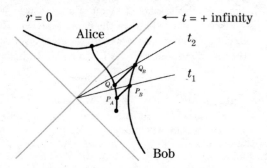

Figure 8: What Bob sees of Alice when it is Alice who is falling into the black hole. He cannot see her crossing the horizon.

If Alice, in her own frame of reference, passes the horizon, she cannot send a signal to Bob warning him: "Hey Bob, I just passed the horizon," like some people send a postcard to their friends when they pass the equator. Why? Because such a signal would travel on a 45° line in figure 8 and never reach Bob.

Everything that Bob can observe and measure, his entire physical observations, his entire universe, all involve Alice outside the horizon. As far as he knows, her heart slows down as she approaches the horizon. It somehow stops beating, in the sense that each beat takes longer and longer. For him, she all but dies at the horizon – but it happens only at the end of times. He can have no idea about what happens to her past the horizon, because for him it happens "after the end of times." She is indeed doomed then.

One important point should be mentioned: we are considering a very big black hole. Indeed, we assume that the tidal forces at the horizon are negligible. The apparent squashing of Alice at the horizon, viewed by Bob, has nothing to do with tidal forces. It is only a variety of Lorentz contraction[10] already studied in the last lecture (see also volume 3 of TTM). No tidal-force-related deformation happens to Alice at the horizon.

For a collapsed star with the mass of the Sun, the Schwarzschild radius is approximately 3 kilometers, and the tidal forces at the horizon would already be huge. The Sun when it collapses may or may not make a small black hole (see questions/answers session). If it makes a black hole, it will be of the nasty type. On the other hand, at the center of our galaxy, there is a humongous black hole, which is rather mild even close to its horizon – but of course you would be ill-advised to go in and explore.

When Bob watches Alice, he sees her slow down. Even her heartbeats slow down – rather zen for someone about to fall into a black hole! Does this mean that Alice sees something special happening to Bob when she watches him?

Let's turn to that question: while she is falling into the black hole, what does Alice see when she looks at Bob? Again all the answers are in the diagrams. Now it is figure 9 that tells the story.

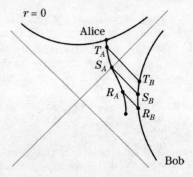

Figure 9: What Alice (falling into the black hole) sees of Bob. She sees him all along – as long as she is alive.

As Alice sails along her trajectory, there is no problem for her to see light rays emitted by Bob at times in her past. When she

---

[10]Also known as Lorentz–FitzGerald contraction.

is at R_A, outside the horizon, she sees Bob at R_B. When she crosses the horizon at S_A, she sees Bob at S_B. Until she hits the singularity at the center of the black hole, in the vicinity of which tidal forces will shred her apart, and where she will finally get annihilated, she continues for a while to see Bob without difficulty. Moreover there is nothing indicating that she would clearly see Bob speed up, or slow down.

In short, Alice can still see Bob when Bob can no longer see her. And Alice doesn't see anything peculiar happening to Bob. To her, everything looks normal and as usual. Of course, Alice cannot see Bob past a point a little after T_B on Bob's trajectory. But that is only because, by then, the tidal forces will have torn her apart and terminated her.

To conclude, there is a total asymmetry between Bob and Alice.

Exercise 1: Use the various diagrams we drew to describe what a third person, say Charlie, following Alice sometime behind her would see of Alice and of Bob at different times.

In exercise 1, so long as Charlie or Alice is still outside the black hole, Bob can receive a message from him or her. And, for Bob, the "last" messages will be infinitely late in time. Thus he does not see either of them ever cross the horizon.

With the diagram, you can also analyze how Alice and Charlie are able to communicate with each other. In the case that Alice and Charlie are both in free fall, their trajectories will be parallel and look almost like straight lines. However they fall, you can think of Bob being eventually unable to see them as a consequence of the fact that he has to accelerate to stay outside the black hole.

In all cases, what one of the protagonists can see of the others is simply and strictly related to the pair of points, in their respective trajectories, that 45° light rays can link.

In short, the diagram says it all.

Questions/Answers Session

This book stems from a course that was given live in front of an audience of students. There were questions/answers sessions to make breaks from the formal lecturing and to offer students an opportunity to get answers to specific questions they had. The questions treated were naturally less organized than the sections of a lecture, yet they will illuminate and complement what we have explained. Here is the first questions/answers session.

Question: Is there a way to figure out how a person would perceive the world around them while falling inside a black hole?

Answer: There are some movies that purport to show what is happening when someone falls into a black hole. Andrew Hamilton has made some simulations of what it looks and feels like to fall into a black hole, which can be viewed on *YouTube*.[11] In my opinion, these movies are not very illuminating. You won't get much out of them, but it might be fun to watch one. They really are disorienting. If you are in an auditorium and one of them is projected on a big screen in front of you, it can make you nauseous and seasick.

Q: If, as you said, when we fall into a black hole, we perceive nothing special happening while crossing the horizon, then what is special inside the black hole?

A: We said that nothing special happens at the horizon but of course the light that is coming at you is coming in very peculiar and funny ways.

You might wonder what happens at the origin in the diagram of figure 5, that is, at the point H on the horizon. Nothing much happens there because, in fact, for a real black hole, which is formed for example by stellar collapse or something like that, the center part of the diagram isn't even on the figure. Only a portion of the diagram in the upper and right parts is relevant.

---

[11] See Andrew Hamilton, "Journey into a realistic black hole," at https://www.youtube.com/watch?v=HuCJ8s_xMnI

Figure 5 in truth is the diagram of a very idealized black hole. For a real black hole that forms by collapse, only a portion of the diagram, which does not include H, means anything.

Q: When things fall into a black hole, the mass of the black hole increases. What happens to its Schwarzschild radius?

A: As a black hole gulps a mass coming from the outside, its own mass of course gets larger. Therefore the Schwarzschild radius gets larger too; see equation (2).

The pictures we drew are appropriate only when the objects or people falling in are much lighter than the black hole. So the black hole doesn't react strongly to them.

If a big mass, of a size not negligible compared to that of the black hole, falls into it, as the mass approaches the black hole, the horizon of the black hole will bulge to merge with the coming mass, as shown in figure 10.

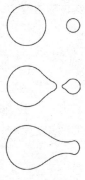

Figure 10: Black hole (shown on the left) gulping another sizable mass (shown on the right).

Then because black holes have no hair, it will quickly pull itself together back into a sphere. The process is extremely fast, of the order of magnitude of light going from one black hole to another. For two solar masses merging, it would be, I believe, of the order of magnitude of a millisecond.

The sphere will be slightly larger. If we know how much mass is added, the dynamic of the process in figure 10 can be calculated with Einstein field equations, which we will study in lecture 9.

Q: How can we see the black hole at the center of the Milky Way?

A: We don't "see" the black hole itself. We identify phenomena that are due to the presence of the black hole.

The light that we see from a black hole is not coming from the horizon. It is coming from all sorts of stuff, hot stuff, that is circulating around the horizon and that is heated up by energy and collisions.

Remember from lecture 6 that at a certain distance around the black hole, for instance, we can have photons of light moving around in circles.

Around the black hole all sorts of very complicated collisions can take place, which send out particles and light, some of which end up in our eyes or measuring apparatus.

In figure 5, we are talking about the idealized situation of one black hole in empty space. But in reality there are all kinds of stuff around it. There is an accretion disk. There are plenty of things falling in.

What you see is not coming precisely from the horizon. It is coming from all this activity at some distance from the horizon.

Q: You said that the merging of two black holes happens very quickly, but you also said that it is related to their sizes. So can't it take very long in some cases?

A: Well in theory, you can consider black holes as big as you like. Then their merging can also take any time you like.

The biggest black hole that is known, however, has a mass of something like 10^9 solar masses. With two such black holes, the merging time we are talking about, the time for them to equilibrate into one perfect sphere, would be of the order of 20 minutes.

Q: Suppose that Alice, while falling into the black hole, carries with her a big mass. Wouldn't that alter the analysis we just made of the communications between her and Bob?

A: Your question really is about what happens inside the black hole when its mass increases. We haven't talked about that, i.e., what happens inside the black hole when a mass is absorbed. This is in truth a question about their formation.

In the next lecture, we are going to describe exactly how a black hole can form, how the horizon forms, and, if you throw in more material, how the horizon responds. We will examine a simple example and see that things are different from expected. They are surprising and yet logical.

Q: Does anything change if Bob is orbiting the black hole?

A: No, not much. It becomes a more complicated problem to be quantitative about exactly what he sees. But there is no fundamental change.

Q: Don't the analyses presented depend on the distribution of mass inside the black hole?

A: No, no. I understand what you mean, but you have to realize that what you call "the distribution of mass" is neither on the horizon, nor nicely spread inside.

There are not even shells of mass inside the black hole – like there is for instance when we go inside the Earth and, because of a theorem due to Newton, which we will talk about in the next lecture and again in the next volume of TTM on cosmology, the gravity goes down to zero as we approach the center. In fact there is no relevant distribution of mass in a black hole. It is all at the singularity, all of it.

One important point to keep in mind about black holes is this:

A black hole is the analog in general relativity of a point mass in Newtonian physics.

All the mass is at the center. There is no such thing as getting deeper and deeper into the mass of the black hole after we pass the horizon, like we can enter into Earth and gravity decreases progressively to zero as we get to the center.

Concerning the *evolution* of the black hole, *how it forms*, etc., these are very interesting questions. We will talk about them, about the creation of a black hole, in the next lecture.

Q: You haven't talked about the angular momentum of black holes. You just said that when they rotate they are no longer spherical. Can you say more about this?

A: The vast majority of things or collections of things in the universe have some angular momentum. As a consequence the vast majority of black holes rotate, and even rotate fast. Why? Because in the process of collapsing and forming a black hole, even if at first the material does not show much rotation, it has some angular momentum. Then, as its dimension shrinks, like an ice skater pulling the arms closer to the body, it will spin faster and faster. This, however, doesn't change the angular momentum, of course. The angular momentum of an isolated system is a conserved quantity.

Rotating black holes are more complicated to analyze than non-rotating ones. The geometry of space-time is quite a bit more involved. They sort of carry some space, as well as some time, along with them in their rotation.[12] The singularity of a rotating black hole is also a different thing from that of a non-rotating one. We won't discuss rotating black holes.

---

[12]On a subject close to rotating black holes, the reader may look up for instance Tipler's cylinders, which can deform space-time in very odd ways, leading to apparent paradoxes like time travel.

We study non-rotating black holes in this course not so much because they are a subject in themselves – and in fact we study only
simple ideal ones – but because they are a natural intermediate
step in the study of general relativity. In the same way, the study
of point masses and springs is a natural intermediate step in the
study of Newtonian mechanics.

Q: Do black holes appear in a process of condensation of existing
matter, or do they have a quantum origin?

A: They don't have a quantum origin. They appear at the end of
the life of certain stars.

Stars form when materials agglomerate and start to radiate. The
interior of the star burns hydrogen or helium or whatever during
its lifetime. When the star eventually runs out of fuel, the forces –
due to the radiation pressure that is preventing it from collapsing
– disappear and the star begins to contract.

Depending on how heavy it is, a star might contract into a white
dwarf, which is a more-or-less ordinary thing, pretty dense but
still made out of nuclei and atoms and so forth. Or if it is heavier,
it might collapse into a neutron star. That is a very compact object, but it can still support itself because the material is strong
enough to support itself against gravity. If it is yet heavier, then
it may contract past its own Schwarzschild radius. Once it falls
past its own Schwarzschild radius, it becomes a black hole.

The Sun, when it eventually "turns off," will not form by itself
a black hole. If nothing else happens to it, like merging with
something else, it will form a white dwarf.

Q: Is it true that black holes can evaporate?

A: In theory black holes can "evaporate" and disappear. But it
is only a possibility stemming from the equations. The process
would take immensely long. It would be faster for small black
holes and slower for big black holes. For instance, a black hole

with the mass of Mount Everest would take as long as about the age of the universe to evaporate.

It is very difficult, however, to imagine such a process in reality. In an object the size of Mount Everest, there is just not enough matter for it to collapse and form a black hole.

The whole Earth could not form a black hole. It is not heavy enough to do anything when it cools down. As already said, the Sun will not form by itself a black hole. At least that's what most people think. It'll form a white dwarf.

There are only two ways we can think of that can lead to the formation of a black hole:

- One is gravity collapse of a massive body as we described.

- The other is via violent collisions. Velocity can replace gravity and slam stuff together. If there is enough violence, a black hole can form even though there was not enough material for gravity collapse.

Q: Is it conceivable that this second type of formation of a black hole could happen in a particle collider?

A: The smallest imaginable black hole that makes any sense is a black hole of Planck mass.[13] By slamming particles hard enough, you could theoretically make small black holes.

Similar collisions happen naturally, with higher energies than in man-made accelerators, when cosmic rays hit the Earth's upper atmosphere. Therefore, possible artificial black holes, created for instance at the LHC in Geneva, would be no more dangerous.

Let's end the questions/answers session here.

The study in more detail of the formation of black holes will be the subject of the next lecture.

---

[13] $m_\mathrm{P} = \sqrt{\frac{\hbar c}{G}} \approx 1.22 \times 10^{19}$ GeV/$c^2 \approx 2.18 \times 10^{-8}$ kg

Lecture 8: Formation of a Black Hole

Lenny and Andy are in a meditative mood. It is surprising how the physics of general relativity, black holes and wormholes, which are high-level math and abstract physics, bring one close to time immemorial questions about the world, the time, the meaning of life, usually the realm of philosophers and ecclesiastics.

Andy: *I'm keen to know more about wormholes. Then I will sneak into one and go say hello to a few of my stars in history.*

Lenny: *I'd love to have a conversation with Albert, or Ludwig, or Joseph-Louis. But I'm afraid, Andy, wormholes are only odd mathematical solutions to the equations. That doesn't automatically bestow on them existence. In fact, we will see why they are only science fiction.*[1]

Introduction

We begin this lecture by recalling what we already learned on black holes: the horizon, the locally flat space-time, the change of variables to Kruskal coordinates – also called Kruskal–Szekeres[2] coordinates – the fundamental diagram of the geometry of space-time near a black hole, what free-falling people and people staying at fixed distances from the black hole experience.

---

[1] They are science fiction insofar as for human beings traveling into the past or other universes is impossible. But they do play a role in quantum gravitation.

[2] George Szekeres (1911–2005), Hungarian-Australian mathematician.

These at first intricate notions should progressively become clearer
and more familiar. The reader should begin to see their deep simplicity. It is imperative to understand them well and feel at ease
with them in order to go further, and finally reach a good understanding of the theory of general relativity, in particular the
Einstein field equations, which will be the subject of the next lecture.

After the reminder, we shall use a different type of diagram to
represent the geometry of space-time near a black hole, and see
what other phenomena it reveals. Finally we shall examine the
formation of real black holes.

Kruskal–Szekeres Coordinates

Last lecture we presented the fundamental diagram of space-time
geometry created by a black hole, reproduced in figure 1. It is the
geometry viewed in the flat coordinates of a free-falling observer.

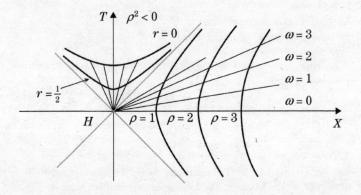

Figure 1: Kruskal diagram near the horizon of a black hole.

Indeed, space-time near the horizon of a black hole is locally, to
a very good approximation, flat space-time[3] in which there is a

---

[3]In fact, it is locally like Minkowski flat space-time anywhere except at
the singularity, just like any smooth surface in Euclidean space looks locally
like a plane. What makes the horizon special however is that it is the radius
of no return.

uniform gravitational field. Someone free-falling near the horizon would perceive space-time as flat. The coordinates X and T used to chart space-time and expressing its local flatness are respectively called *coordinate space* and *coordinate time*.

We built hyperbolic coordinates showing that the Schwarzschild metric of space-time[4] created by a black hole and its gravitational field, is equivalent near the horizon to a flat metric expressed in a uniformly accelerated frame.[5] This is the analog in general relativity of what we saw in lecture 1 in Newtonian mechanics.

Namely, when we are in the ordinary space-time of Newtonian mechanics and there is no gravitation field, say, far away from any gravitating body, then in the frame of reference of a uniformly accelerated elevator, an object – be it fixed or with a certain movement, like for instance a photon of a light ray, or free – appears to be submitted to a uniform gravitational field.

Analogously, near the horizon of a black hole, people staying outside the horizon at fixed distances have the same sensations as if they were uniformly accelerated in a flat space-time.

When we consider this phenomenon, the specificities of the black hole don't even play a role. It is true in the vicinity of any massive body – like for instance the Earth. Indeed, while standing motionless in front of you, I have a similar feeling: although I don't move, I experience gravitation. It is exactly equivalent to the feeling I would have if I were uniformly accelerated in space with no gravitation field. We are so used to being subjected to gravitation, equivalent to being in an accelerated elevator, that we don't even think of it. But astronauts coming back from a long stay at the International Space Station are acutely reminded of this gravitation/acceleration when they come back to Earth.

---

[4]Remember that the metric of a space, or space-time, is an intrinsic feature of that space. It does not depend on the coordinates. But it is expressed in a system of coordinates with the general formula

$$d\tau^2 = -g_{\mu\nu}(X)\ dX^\mu\ dX^\nu$$

In a different system of coordinates, the tensor $g_{\mu\nu}$ would have different components. The rules of transformation of the components of a tensor were studied in lecture 2.

[5]Uniformly accelerated frames in relativity were studied in lecture 4.

Let's go back to the fundamental diagram of figure 1 and recall its main features. The 45° lines represent the motion of light rays either going to the right or going to the left. The right quadrant represents the exterior of the black hole, that is, places farther away than the horizon. The second, also called upper, quadrant represents the interior of the black hole. We saw that not only point H but the entire lines at 45° form the horizon. We are going to see that the left quadrant and the bottom quadrant don't have any physical significance for a real black hole. Only the right quadrant and the upper quadrant have true significance.

Outside the black hole, somebody at rest at a fixed distance r, or equivalently a fixed distance ρ from the horizon,[6] follows in space-time a trajectory that in the (X, T) frame of reference is represented by a hyperbola. As we said, this person is experiencing an effective acceleration.

To nail these important points, let's recall in more detail the geometry of space-time in Newtonian physics.

On the surface of the Earth, in Newtonian physics, we represent the geometry of space and time with a Euclidean space plus a universal time coordinate. For simplicity, we reason with only one spatial axis, which is radial from the center of the Earth.

Various points above each other[7] fixed with respect to Earth, when viewed in the stationary reference frame of an observer at rest on Earth, have trajectories that are parallel lines as time varies (they are the vertical dotted lines in figure 3 of lecture 1). Notice also that lines of fixed time don't converge to a center point but are parallel too, orthogonal to the trajectories. Finally the time coordinate is the same in any reference frame. It is called the universal time.

A uniformly accelerated frame is simply that of an observer inside an accelerated elevator.

---

[6] ρ is the proper distance. When $r > 1$, ρ is related to r by a simple increasing function that we studied in lecture 7.

[7] We say "above each other" because the spatial axis is a radius from the center of the Earth, but on our graphs it is represented as the horizontal axis.

The trajectories of points now fixed with respect to the accelerating elevator, when they are viewed in the stationary frame of the observer at rest on Earth, are parabolas (figure 4 of lecture 1). We would of course have the same situation if we represented, in the frame of the elevator, points fixed with respect to Earth.

Continuing to nail the important points, let's turn to space-time in the theory of relativity and again let's go into more details.

In relativity, the geometry of space-time is different from that in Newtonian physics. Space and time are intimately mingled. Simultaneity, for instance, is frame-dependent. However, a uniformly accelerated reference frame, far away from the origin of the Minkowski diagram, is very much like an ordinary accelerated reference frame near the surface of the Earth (see figure 15 of lecture 4): trajectories are almost parallel vertical lines, and coordinate time and proper time are almost the same.

In figure 1, spread on a radial axis jutting out of the horizon of the black hole (angles θ and ϕ play no role, they are constant), we consider a collection of people, at proper distances respectively $\rho = 1$, $\rho = 2$, $\rho = 3$, etc. When the time variable $\omega = 0$, these people are regularly spaced on the horizontal axis. Then, as ω increases, the axis they are on turns while still going through the point H. The people are still regularly spaced, and this even remains true on the fundamental diagram showing their hyperbolic trajectories.

Coordinates T and X, in figure 1, are those of a free-falling observer, who is anywhere in space-time except at the singularity $r = 0$, and happens to look, in their coordinate frame, at what is going on near the horizon of the black hole. The observer sees the collection of people at the various ρ's as being uniformly accelerated. They form what we *defined* as a uniformly accelerated frame in relativity; see lecture 4.

The actual acceleration – the physical push felt by each person – is not the same at different ρ's. The smaller ρ is, the stronger is the acceleration. This is shown by the growing acuteness of the hyperbolas near H. On the other hand, the acceleration stays the same along a hyperbola of constant ρ.

In the right quadrant, the straight lines fanning out of H are the lines of constant proper time t or ω for the people in that accelerated frame. As ω tends to $+\infty$, the lines become closer and closer to the 45° line of a light ray emitted from H.

We just described the outside of the black hole, that is, what happens in the right quadrant. By contrast, the inside of the black hole is represented in the upper quadrant. In this upper quadrant, there is a point where the geometry of space-time is no longer equivalent to locally flat. There is something nasty happening. It is the singularity at the center of the black hole. Strangely enough, the singularity is represented by a whole curve in figure 1. It is the hyperbola marked with $r = 0$. The tidal forces are very strong near the singularity.

The problem for an adventurous person exploring space-time is that once you find yourself inside the horizon of a black hole, i.e., in the upper quadrant, there is no way out – unless you could travel faster than c. Since your trajectory tangents are always steeper than 45°, you will inescapably eventually hit the singularity, and that will happen in a finite amount of your own time.

In summary, outside a black hole people, objects, or light rays can enter into the black hole. They also have the possibility to stay outside. But once inside the black hole nothing can get out, and everything ends up at the singularity in a finite amount of proper time. That is all there is to know about the geometry of black holes. Although outside and inside it is locally approximately equivalent to flat space-time, it is not so everywhere. There is a singularity. Furthermore the horizon is the distance of no return.

Penrose Diagrams

It is very convenient to redraw the diagram in figure 1 differently. We are going to end up with Penrose coordinates and Penrose diagrams. But these are built in several steps. As usual, let's take the steps one by one.

In the geometry of space-time created by a black hole, space is infinite. Time may come to an end at the singularity, but the lightlike directions off at 45° are also infinite. Therefore, in figure 1, we cannot draw the entire space-time in the limited dimensions of the page. For many purposes it would be useful, however, to be able to draw the entire space-time on a finite sketch. We will do it, and it will provide some new visual tools and sustain intuition.

Let's try doing that, but first of all let's do it with good old flat space-time. We shall do a coordinate transformation that pulls the whole thing into some finite region of the page.

This is useful, incidentally, for geometries in space-time that display rotational symmetry. Rotational symmetry, or invariance, means that what happens at any event in space-time doesn't depend on the direction of the event from some center.

Let's focus for a moment on ordinary 3D Euclidean space. To describe a system that displays rotational symmetry, it is usually useful to use ordinary polar coordinates. And when we have rotational symmetry, we often don't need to worry too much about the angular direction. Aside from time, the other coordinate that matters is the radial distance.

Therefore, to start with, before shrinking the whole space-time into the page, we think of ordinary flat space-time as having a time axis T and a space axis R, which is a radial direction, as shown in figure 2. The time goes from $-\infty$ to $+\infty$. And the spatial coordinate R goes from 0 to $+\infty$.

We put some markers: $T = 0$, $T = 1$, $T = 2$, $T = -1$, $T = -2$, etc. And $R = 0$, $R = 1$, $R = 2$, $R = 3$, etc. The entire space-time is not yet shrunk on the page. So far T can go up to heaven, and down to hell. Similarly, the radial axis R doesn't have any limit on the right.

In all cases, we will imagine using units so that the speed of light is $c = 1$. For instance, if we used seconds for time, then the space units would be light-seconds.

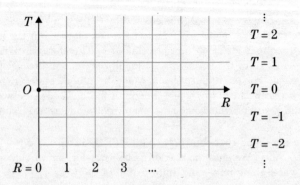

Figure 2: Flat space-time shown with the most ordinary diagram.

Then some light rays follow 45° lines. But not all of them. Light rays coming from the past *aimed at the spatial origin* and then going on into the future, form a pair of 45° half lines, as shown in figure 3.

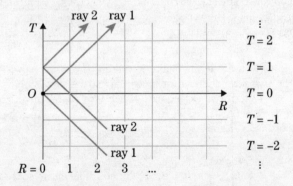

Figure 3: Two light rays aimed at the spatial origin.

Another picture will help us understand why some light rays follow paths like in figure 3, and others don't. Figure 4 shows space in polar coordinates centered on the same point as above. The time axis is not represented.

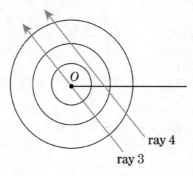

Figure 4: Two light rays, one aimed at the center, the other not.

In figure 4, ray 3 is aimed at the center. That ray, represented in figure 3, would appear to bounce against the vertical line $R = 0$, like ray 1 and ray 2.

But ray 4, if we represented it in figure 3, would not bounce against $R = 0$. It would be coming from far away in the past, from what we call *light-like infinity*. As long as it is far away, unless we have very good instruments, we could not say whether it would hit the center. But near the vertical line $R = 0$, it would swerve before hitting it, as shown in figure 5.

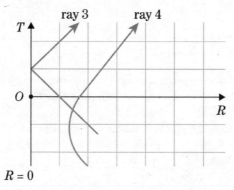

Figure 5: Light ray 4 passing by the origin, but not hitting it.

Light ray 4 doesn't hit the origin because it was not aimed at it. Far away from the small radii it follows almost 45° lines in figure 5. In fact the trajectory of ray 4 is a hyperbola.

In figure 5, light ray 4 looks like it is repelled from the origin. Of course, it is not repelled from anything, it just passes by near the origin before flying off again, all in a straight line as shown in figure 4. The phenomenon appearing in figure 5 really is what we call centrifugal force. It is the centrifugal force that keeps the light ray from hitting the origin.

We went over those considerations in figures 2 to 5 to establish some preliminaries and prepare for the next step.

The next step is to squeeze the diagram in figures 2, 3, and 5 to fit the entire space-time on the page. In particular, what we dubbed the light-like infinity will now appear as a point somewhere on the graph. Of course, the diagram will be deformed. It is not going to look the same. But we are going to keep one feature fixed: light rays will keep following 45° lines or asymptotes. That is a useful thing to do because then we can see how light rays move, and we can see what is going slower or faster than the speed of light.

The shrinking of the whole space-time into a limited diagram is done mathematically itself in two steps. We introduce a first set of new coordinates:

$$T^+ = T + R$$
$$T^- = T - R$$

$$(1)$$

In this set of coordinates, lines of constant T^+, for instance, are diagonal lines at $-45°$ angle, as shown in figure 6.

Similarly, lines of constant T^- are diagonal lines at $+45°$. Thus now we have two sets of coordinates to describe our flat space-time. We have (T, R) and we have (T^+, T^-).

Let's look at the vertical line going through O. It is $R = 0$ in the old coordinates (T, R). And it is easy to see that in the new coordinates (T^+, T^-), it has the equation

$$T^+ = T^-$$

$$(2)$$

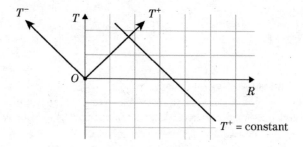

Figure 6: Coordinates $T^+ = T + R$ and $T^- = T - R$.

One way to see it is to notice that it is the usual equation for the line splitting the angle formed by the two axes T^+ and T^- in half. Or we can write $T^+ = T^-$ as $T + R = T - R$, which becomes $2R = 0$, or $R = 0$. For obvious reasons, the coordinates (T^+, T^-) are called *light-like coordinates*.

Now we introduce a second set of new coordinates, so as to shrink the whole plane (T^+, T^-) into a bounded area of the page. The second set is U^+ and U^-. The coordinate U^+ will be an increasing function of T^+ such that when T^+ goes from $-\infty$ to $+\infty$, the coordinate U^+ goes from -1 to $+1$; see figure 7.

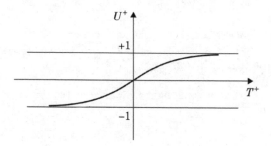

Figure 7: U^+ as a function of T^+.

There are plenty of such functions. It is customary to use the hyperbolic tangent. We will also apply the same transformation to T^-. Thus we define U^+ and U^- as follows:

$$U^+ = \tanh T^+$$
$$U^- = \tanh T^-$$

(3)

You don't need to know much about the hyperbolic tangent, except that is has a graph as in figure 7 (we have taken a horizontal unit twice as long as the vertical one in the figure).

Now we represent the flat space-time of figures 2 to 6 with coordinates (U^+, U^-). When T^+ goes to infinity, U^+ never gets bigger than one. Same for T^- and U^-. It becomes figure 8:

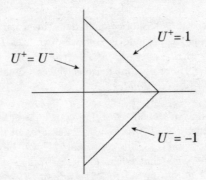

Figure 8: Space-time in coordinates (U^+, U^-).

All we've done is squish the geometry of figure 6, vertically and horizontally, onto a finite triangle. This will become clearer when we look at light rays in figure 9.

Let's look at an in-going light ray aimed at the origin. It corresponds to

$$T^+ = \text{constant}$$

After hitting the origin, it becomes an outgoing light ray from the origin. It then corresponds to

$$T^- = \text{constant}$$

The coordinate transformation using tanh has the property that such light rays are still straight lines as shown in figure 9.

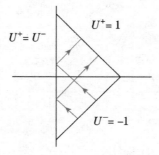

Figure 9: Light rays going through the origin, in (U^+, U^-) coordinates.

Another enlightening graph is that of constant times; see figure 10.

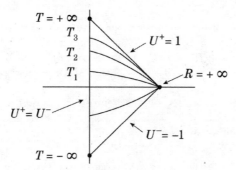

Figure 10: Lines of constant time.

We have also squished "spatial infinity" into the point $R_{+\infty}$. It is called *space-like infinity*. As we look at bigger and bigger fixed times, the corresponding lines, when R increases, will be closer and closer to the line $U^+ = 1$.

Now let's look at fixed spatial positions. The corresponding trajectories will all go to $T_{+\infty}$, coming from $T_{-\infty}$. We see them in figure 11.

The diagram is "the same" as that of figure 6. But we have operated a change of coordinates so that every point that had some coordinates (T, R) gets mapped somewhere onto a finite triangle.

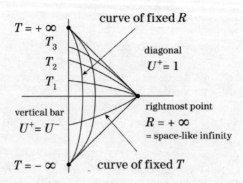

Figure 11: Fixed T's and fixed R's in (U^+, U^-) coordinates. The fixed R's naturally come from $T_{-\infty}$ and go to $T_{+\infty}$.

The process of bringing everything from figure 6 into a finite diagram (figure 11), without changing the way radial light ray trajectories appear (figure 9), is called *compactification*. Figure 11 is called a *Penrose*[8] *diagram* or a *Carter*[9]*-Penrose diagram*.

That closes our section on the Penrose diagram of a flat space-time. Now we turn to black hole geometry. This will lead us to a discussion of wormholes.

Wormholes

We succeeded in representing the entire flat space-time on the page. Now the question is: can we apply the technique to black hole geometry?

To start with, let's recall some terminology: the right tip of the triangle in figure 11, where $R = +\infty$, is called spatial infinity, or space-like infinity.

Let's introduce some new natural terms: the tip of the triangle on top, along the right side, where $T = +\infty$, is called *future time-like infinity*. And the tip at the bottom is called *past time-like infinity*.

---

[8]Roger Penrose (born 1931), English mathematician and physicist.
[9]Brandon Carter (born 1942), Australian theoretical physicist.

Furthermore, the 45° line between $T_{-\infty}$ and $R_{+\infty}$ is where all light rays come from, as in figure 9. And the line between $T_{+\infty}$ and $R_{+\infty}$ is where they all go, after hitting the origin or swerving past it.

The standard notation for the segment $T_{-\infty}$ to $R_{+\infty}$ is $\mathcal{I}^-$, read "script i minus," or sometimes "scry minus." And the segment $T_{+\infty}$ to $R_{+\infty}$ is $\mathcal{I}^+$, read "script i plus," or "scry plus."

They are called respectively *past light-like infinity* and *future light-like infinity*. They are the places where light rays begin and where light rays end. This is all summarized in figure 12, which shows what *flat space-time* looks like when compactified.

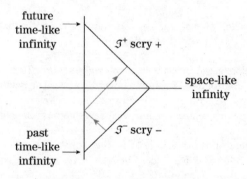

Figure 12: Flat space-time compactified.

Let's turn our attention to a *black hole*. How do these diagrams apply to the geometry of space-time created by a black hole? We can do exactly the same thing. We can take the entire diagram in figure 1 (where we can rename the X-axis the R-axis), that is, the original Kruskal coordinates and diagram. We can again introduce light-like coordinates, $T^+ = T + R$ and $T^- = T - R$, and shrink them into functions varying between only -1 and $+1$. We do exactly the same operations. What shall we get?

What it will look like is the diagram in figure 13. Again it shrinks the original four quadrants into four squares, but the upper square is halved by the singularity. And we disregard momentarily the lower quadrant.

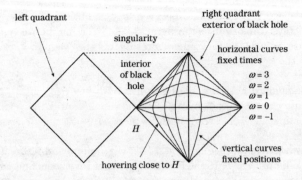

Figure 13: Black hole space-time compactified into a Penrose diagram.

Note that the letter H, for horizon, denotes the entire line slanted at $+45°$ in the middle of the diagram.

The exterior of the black hole is the right square. In this square, the hyperbolas of fixed positions hovering above the horizon H become the curves going from bottom to top. The original straight lines of constant time ω become the curves going from left to right.

The interior of the black hole is the upper half square. It is the place where, if we find ourselves in it, we are doomed. It is bounded by the singularity line. The singularity is not off at infinity. Remember that once we are inside the horizon, we cannot stay at a fixed point on the graph because we cannot avoid the future,[10] and it takes a finite amount of proper time to travel to the singularity and be annihilated by infinite tidal forces.

Figure 13 shows the entire space-time of a black hole in a Penrose diagram. If we are in the right part, we can stay where we are, but that takes acceleration in the sense that we have to oppose gravitation pulling us in. We can even escape to farther away. We can also fall beyond the horizon like through a floor.

---

[10]Not being able "to avoid the future" is also true in the right square, of course. But there, even if we stay at a fixed position in space, our trajectory in space-time will go to infinite time (its top) while keeping away from the inside of the black hole. Therefore we are never destroyed by the singularity.

The picture in figure 13 raises an obvious question: what could possibly correspond to the left square? Same for the lower part. These are the compactified representations of the left quadrant and the lower quadrant of figure 1.[11]

The left quadrant and the lower quadrant have no real meaning for a real physical black hole. We will see that when we work out a real physical black hole, how it forms, etc.

Nevertheless, we can ask ourselves: what kind of geometry is described by the full extended Kruskal–Penrose diagram of figure 13? It seems to have two exterior regions – the right and left squares – connected together at the horizon H. Remember that, at the horizon, $R = 2MG$. This value of R, denoted R_S, is called the Schwarzschild radius. It is the place where $(1 - 2MG/R)$ changes sign. We arrive from the right, that is, from the exterior region, the coefficient changes sign, and we continue in the interior region, represented in the upper quadrant or upper half square.

Now suppose, at a fixed time, say $T = 0$, we look at a slice of space, or at the entire space. What does it look like? In figure 13, it is just the complete horizontal segment from the right to the left of the figure. But in all our diagrams so far, for the sake of drawing, space was only one-dimensional. Think of space as three-dimensional. Then for each R, it is the surface of a sphere – also called a 2-sphere – of radius R.

When we start far away, the celestial sphere is very big. Then, as we approach H, the 2-sphere shrinks to $R = 2MG$.

If, for drawing purposes, we think of space as only two-dimensional, for each R the 2-sphere becomes a "1-disk," that is, the boundary of a disk, in other words a circle. Flipping the diagram from horizontal to vertical with H down below, we can represent these circles in figure 14.

---

[11] A bit of epistemology: the reader may think that these speculations are very abstract and a kind of game with pictures and mathematics. But in truth, that's how the human brain reflects on *any* perception. Our senses provide us only with raw perceptions, which the brain organizes into phenomena described using a 3D space plus time. The theory of relativity, which at first left most people skeptical, made these interpretations more involved.

On figure 13, at the horizon H, instead of turning into the interior region (upper half quadrant), let's suppose that we just "kept going" left. It is represented as the lower part in the flipped figure 14. It corresponds to the left quadrant in the previous Penrose diagram.

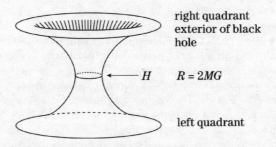

right quadrant
exterior of black
hole

$\leftarrow H$ $R = 2MG$

left quadrant

Figure 14: Wormhole.

Once we have passed the horizon in this way, the space of each R starts expanding again. It looks like we can pass from one side to the other by going through what people call a *wormhole*. It is also called an *Einstein–Rosen*[12] *bridge*.

It connects two seemingly external regions of the black hole, which get bigger and bigger as we move away from the bottleneck. In other words, it looks like the black hole is connecting two universes, or two asymptotic regions.

You might think: well, we could pass through the bottleneck at the horizon going from right to left. But we can't. Let's think of somebody who wants to make such a trip. If that person starts anywhere on the right and is to end up anywhere on the left, their trajectory at some point must have a slope smaller than 45°, that is, they will have to exceed the speed of light.

---

[12]Nathan Rosen (1909–1995), American Israeli physicist who in the 1930s collaborated with Einstein in general relativity on these "bridges," now more commonly called wormholes. He also collaborated on the so-called EPR (Einstein–Podolsky–Rosen) paradox in quantum mechanics.

In fact the diagram of figure 14 is somewhat misleading. It shows everything *at an instant of time*. But there simply isn't really time to go from the upper part to the lower part. We cannot go from the real exterior of the black hole to the left quadrant (shown in the lower part of figure 14). Therefore the Einstein–Rosen bridge isn't really a bridge.

One way to think about it is that the bottleneck opens up and closes again before anything can pass through it. But the best way is just to look at figure 13 and say: "yeah, if we could exceed the speed of light and move horizontally, yes we would move from right to left going through the neck at the center. However we can't do that. We can only move along 45° lines or steeper."

Wormholes have been the source of much science fiction, passing from one universe to another through the horizon of a black hole (not into its inside but into its left region). However as we saw, it cannot happen. These wormholes don't allow us to access other universes. They are of a kind called *non-traversable wormholes*. There are also versions where we could apparently travel into the past, but they are as fantastic as those we just described.

It may come as a disappointment that we won't be able to go on a field trip to other universes or go say hello to Moses at the end of the lecture. But in fact, as we shall see in the next section, there is no real meaning to the left side of figure 13 anyway. It is not a real place.

Let's turn to the creation of a black hole, not in the laboratory because that is too hard, but in an infinite space by having some in-falling matter.

Formation of a Black Hole
and Newton's Shell Theorem

In order to study the formation of a black hole, we are going to take a very special kind of in-falling matter.

We begin with a point in space-time, where there is no black hole nor anything, and with a very distant shell; see figure 15. The

shell is not made of iron or other matter like that. It is a thin
shell of incoming radiations.

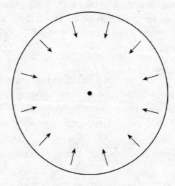

Figure 15: Shell of incoming radiations toward a central point.

Radiation carries energy. Radiation carries momentum. We suppose that, for one reason or another, it has been created far away.
And it is incoming, with spherical symmetry, at the speed of light,
toward a central point.

At some point in time for the observer, as the shell radius gets
small enough, there will be so much energy in a small region that
a black hole will form. This is the simplest version of black hole
formation.

A star isn't really made up of things that fall in with the speed of
light. What we are looking at is the idealized problem of a *black
hole* being formed by stuff coming in with the speed of light, the
stuff being on a thin spherical shell. It is also one of the simplest
relativity problems.

There are only two important things to know to understand this
formation of a black hole:

1. The shell moves in with the speed of light.

2. We need to know a famous theorem of Newton in classical
   mechanics, and a version of it in relativity.

Here is what Newton's theorem in classical mechanics says. If we have a shell of matter forming a 2-sphere, that is, uniformly spread on the surface of a sphere, then the following hold:

(a) The gravitational field *inside* the sphere is null.

(b) The gravitational field *outside* the sphere is identical to that of a point mass at the center with the same amount of mass.

It is even true if the shell is moving, for example, collapsing toward the center. Of course, as time passes, there will be less inside space, with no gravitational field, and more outside space, with a gravitational field. But it is still true.

The theorem is also true in general relativity. It now says the following: if you have a shell of in-falling mass or energy of any kind, then the interior region is just flat space-time. It is like space-time where there is no point source, or no mass.

Thus, for the interior of the shell, the theorem is similar to its classical mechanics version.

For the exterior region, the situation is different. It doesn't look like a Newtonian point source with an ordinary gravitational metric around it. The reason is simply because no such thing exists in general relativity. What does it look like then? It looks like the Schwarzschild metric, that is, the solution to Einstein field equations that we shall study in lecture 9.

In other words, in general relativity the *inside* of a shell of mass or energy (it's the same thing) is flat space-time. And the *outside* of it looks like a Schwarzschild black hole.

If you have a static, non-moving shell, what you would do to construct the actual solution, i.e., the actual metric, is sort of paste together the metric inside, which is flat, with the metric outside, which is the Schwarzschild metric.

Notice that we do the same thing in Newtonian physics: we paste together no gravitational field on the inside with, on the outside, the standard gravitational field due to a central point source. And it is the way we solve problems of Newtonian physics involving a 2-sphere of mass.[13] We will do exactly the same thing in general relativity.

Let's redraw, with fewer details, the Penrose diagram of a black hole to illustrate what we are going to find; see figure 16.

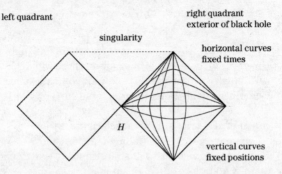

Figure 16: Penrose diagram of a black hole.

And let's also redraw the Penrose diagram of flat space-time. But we now add an incoming shell of radiations, shown in figure 17. We can think of the incoming shell of radiations as a sort of pulse of incoming photons, a pulse distributed nicely on a sphere.

The pulse of incoming photons comes from past light-like infinity (what we also call scry−, pronounced "scry minus"). In 3D they form a continuous collection of shells focusing on the origin, but in figure 17 where we can draw only one spatial dimension, they are represented simply as a straight line. Nonetheless, let's think of it as a shell.

At any given instant of time, where is the interior and where is the exterior of the shell? The diagram makes it easy to answer. The triangle represents space-time with R and T. But both shrunk so that the entire half plane of figure 2 fits into the triangle.

---

[13]Notice a comparable fact in electromagnetism: consider a conducting body with electric charges in a stable, that is static, configuration. Then − whatever the shape of the body − the charges are on the outer surface, and there is no electric field inside the body.

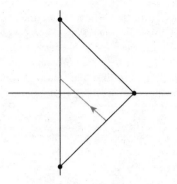

Figure 17: Pulse of incoming photons in flat space.

A point in time and space (the latter being represented with only one dimension) is a point in the triangle, as in figure 18. The trajectory of the shell is the straight line (parallel to the top side of the triangle). All the events at a given time form the curve going to space-like infinity on the right (see also figure 11). At a given time, the shell is at the intersection. The interior of the shell is the part of the curve to the left of the shell. The exterior goes from the shell to space-like infinity.

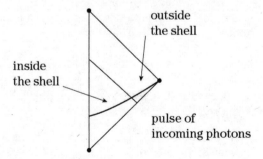

Figure 18: Interior and exterior of the shell at any given instant of time. The curve is a given time. Where it intersects the shell is the position of the shell at that time.

What Newton's theorem or its general relativity version says is that on the interior of a shell everything is flat space-time.

On the diagram of figure 18, we have a dynamic view of the interior of the shell as it moves. In other words, it is correctly represented by the space-time that we drew in figures 8 and 9, which was just a representation of flat space-time. Therefore, first of all, for the interior of the shell the Penrose diagram of flat space-time is the correct representation of everything that is going on, which is the shaded area of figure 19.

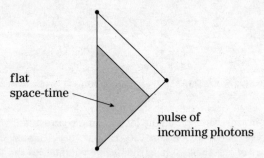

flat space-time

pulse of incoming photons

Figure 19: Flat space-time inside the shell as the shell moves in time.

What about the exterior? On the exterior we are told that there is a gravitational field. And we know that it does not look like flat space-time.[14] Therefore the upper region of the triangle in figure 19, shown in white, is not the correct representation of the physics or of the geometry of the in-falling shell.

Beyond the shell, what is the correct representation? It is the representation of the Schwarzschild black hole of figure 16. Let's redraw it without all the unnecessary details about fixed time curves and fixed position curves; see figure 20.

Somebody on the outside (the right square) throws in the shell. The in-falling shell, represented with one spatial dimension, is a radially incoming light ray. It must move along a 45° straight line. It comes in from far away. It experiences nothing special when it crosses the line called horizon. It keeps going and eventually hits the singularity. That is what the light ray would look like on the Schwarzschild geometry.

---

[14]Recall that a uniform gravitational field is still flat space-time, since by placing ourselves in a "free-falling frame," the Minkowski metric is correct.

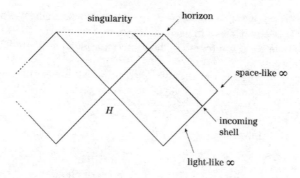

Figure 20: Shell in the black hole diagram, as it moves in time.

Which part of the diagram, in figure 20, is correctly representing the physics that we are doing? Notice that the interior of the shell is not correctly represented by the black hole diagram of figure 20, because its correct representation is the flat space-time diagram of figure 19.

On the other hand, the exterior of the shell is correctly represented by the black hole diagram shown in figure 21, as the shaded area.

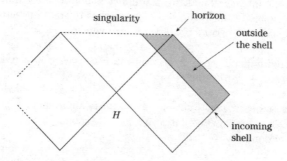

Figure 21: Outside the shell, over time, in the black hole diagram.

We have two diagrams, one representing correctly the inside and incorrectly the outside, and the other representing correctly the outside but incorrectly the inside.

How do we put the two together in order to make a single geometry? It is pretty easy. In each diagram, we throw away the wrong
part, then we paste together the correct parts, as in figure 22.

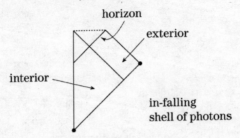

Figure 22: Interior and exterior of the shell pasted together.

The in-falling shell is the thing that the two parts of figure 22
have in common. On one side of the shell, we put flat-space geometry. On the other side, we put black hole geometry. This is
applying Newton's theorem. The complete diagram represents the
geometry of a black hole that is being formed by an in-falling shell.

In figure 22, we also plotted the horizon. It is the 45° line ending
up at the intersection of the singularity and scry+. Nothing special is experienced by someone crossing the horizon. But once you
are inside the horizon, there is no way you can avoid eventually
hitting the singularity. On the other hand, if you are outside the
shell and outside the horizon, you can get away. You can escape
to the scry+, for example.

Notice something very curious. Even in the little shaded triangle
shown in figure 23, where it seems that life should be trouble free,
you are doomed.

In that little triangle you are still in flat space-time. The shell
hasn't even gotten in yet. You cannot see the shell. If you looked
backward on your light cone, you would not see it, because for
you looking backward means looking either at light rays coming
from the past parallel to the shell trajectory or on the other 45°
line of rays coming from the other direction. In both cases, they
don't meet the shell in the past.

You don't know the shell is coming. You are standing in the flat, apparently trouble-free, space-time region. Yet you are also standing behind the horizon defined by the shell. In other words, even though the shell is still far away, and the horizon still very small, you are already in the doomed region.

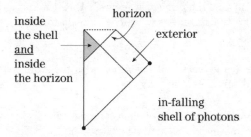

Figure 23: Surprising region inside the shell, where everything is flat and apparently trouble-free, but also inside the horizon. Therefore people there are doomed.

To understand this paradoxical situation, let's see what could happen. If you choose not to move in space, the shell will come in and pass you. Then there is no question that you will be inside the Schwarzschild radius. If you choose to move, trying to get out, you can only move on trajectories steeper than 45°. So you will hit the singularity anyway, and in a finite amount of your proper time at that.

The horizon itself starts very small at some time in space-time (see the next section for a detailed discussion). Then, as the shell approaches, the horizon grows. At some point in time, the shell crosses its horizon. Then the horizon does not grow anymore. In figure 23, the horizon, after the intersection with the shell, corresponds to a fixed $R = 2MG$.

The usefulness of the Penrose diagram, combining the geometries inside and outside the shell, is to show us clearly, given that we cannot exceed the speed of light, that there is no path out from beyond the horizon, no matter where the shell is.

Of course, if we are inside the shell but outside the horizon, we are not doomed. As long as the shell has not passed us, we can stay where we are effortlessly. Once the shell has passed us, to stay still we will have to fight gravitation with acceleration (like you and I do when we are sitting on a chair, even though we tend to forget it). If we don't fight acceleration, we will fall beyond the horizon and face the dire consequences. But if we accelerate, not only can we stay fixed, but we can even fly off to scry+ if we move fast enough.

Let's look at the black hole formation with the more conventional concrete diagram shown in figure 24.

Inside the shell, space-time is flat. Outside the shell, space-time has the same metric as that of a black hole of the same mass at the center of the picture. The horizon is growing toward the Schwarzschild radius, then it doesn't grow anymore. Whatever is inside the horizon is doomed *even if the shell is not there yet.*

So long as the shell has not crossed its horizon, it could presumably change its mind and accelerate back out. Well, we can suppose that shells of photons don't have minds. But if they had one, the photons could turn around as long as they are farther away than $2MG$.

When the shell crosses the horizon, nothing special happens, except that now the photons are trapped for good. They can no longer turn around. Or, well, they could try to turn around, but would not get very far. They are also now doomed to hit the singularity.

In figure 23, we can visualize the point where the shell (the trajectory slanted at $-45°$) crosses its own horizon. It happens at $R = 2MG$. Afterward, we can say that the black hole has formed for good.

Let's discuss the paradoxical aspect of the horizon before the shell has crossed it, using figure 24. It seems that as long as the shell is outside its horizon, someone inside the horizon could possibly move outside of it, while still inside the shell, because in that

region space-time is tranquil. Therefore the person would escape
its fatal destiny. But that is a mistaken view. There would not
be time. Or, if you prefer, the person would have to exceed the
speed of light.

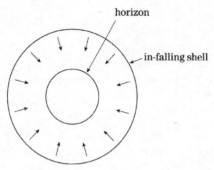

Figure 24: Shell, black hole in formation, and its horizon.

That is the superiority of the diagram of figure 23 over the more
conventional diagram of figure 24. The Penrose diagram is more
abstract, but it shows explicitly the possible and impossible space-
time trajectories.

The horizon is a quaint thing. It begins to exist in a ghostly way,
and at first is very small. That happens at some point in time
after the shell has begun to fall in but is still far away. The horizon grows up to the Schwarzschild radius and then does not grow
anymore. Remember that the horizon is the point of no return; it
is one of its equivalent definitions.

Exercise 1: Suppose that there is a spherical mirror at some
distance farther than the Schwarzschild radius, and that when
the in-falling radiations hit the mirror, they are reflected back
outward.

1. In figure 23, draw the trajectory of the shell of radiations.

2. Discuss in which region people are doomed.

Discussion of the Time Variable

Let us look more closely at the relationship between figure 23 and figure 24, and at the significance of the time variable.

Figure 24 is the conventional picture of the in-falling radiations. *It is a snapshot at an instant of time.*

Figure 23 is the Penrose diagram for the formation of a black hole. *It shows in the same picture space and time, that is, the entire space-time.*

In order to speak of an instant in time, we have to specify which time we are using. Time is just a coordinate. We can make coordinate transformations. To talk about an instant in time really means picking a surface, from a collection of surfaces that are everywhere space-like; see figure 25. From this collection of space-like surfaces indexed by some number, time can be taken to be the index of the surface we are on.

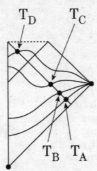

Figure 25: Defining a time variable simply as the index of a collection of space-like surfaces.

If we follow this time variable, at time T_A there is a shell far away. There is no horizon. At time T_B the shell is closer. The horizon has not begun to form yet. At time T_C, the horizon exists, but the radiations are still outside. Sometime between T_C and T_D, the shell crosses its horizon. Then we can say that the black hole exists.

To summarize: as the shell falls in, at some point in time the horizon appears, at first as a tiny point. It grows until the shell meets it, at $R = 2MG$. Then it stays at the Schwarzschild radius. As long as the horizon is below the shell, it is in the flat space-time region. Someone crossing the horizon doesn't feel anything. On either side the person is still in gravitation-free space. However, beyond the horizon, even in the flat region, the person is doomed. The boundary is the upper 45° line in figure 25. It is not when we cross this boundary, that is, the horizon, that we begin to feel gravity. It is when we cross the shell.

The importance of the diagrams cannot be overemphasized. If you try to do calculations to find out where, for example, the horizon forms, you will find that you can't do it. You will wind up drawing the picture, drawing the line from the intersection of scry+ and the singularity, at 45°.

If you want to get a good idea of what is going on, the way to do it is to become familiar with all the diagrams in this lecture, particularly with the Penrose diagrams – for flat space-time, for the black hole, and for their combination.

If you try to reason the problems out, without a good picture in front of you that accurately represents the relationships between the different parts of space-time, you will run into difficulties.

By the way, the important relationship is not the relative size of things in the pictures. A small circle at the intersection of scry+ and the singularity represents a huge space-time region. The same small circle where the horizon is born, along the vertical side in figure 25, represents a tiny region. The important relationships shown by the Penrose diagrams are what light rays – which are 45° lines – and other trajectories can link or not.

The diagrams illustrate *causal relationships*, what can cause what, what signals can propagate, who can send a signal to whom. Can a signal from Alice here get to Bob there? In technical language we would say that the diagrams reflect the causal structure – cause and effect. It means: what can influence what. The rule is that an event E – i.e., a point in the space-time diagram – can only

influence things in front of it, in other words things in the region
where light emitted from E or slower-traveling objects sent from
E can get to. Event E cannot influence events outside its light
cone.

The Penrose diagrams were built first of all to be able to draw and
examine everything on one page, and secondly to reflect faithfully
the causal relations, what can have an effect on what. For that
reason they are constructed in such a way that light rays still
travel along straight lines tilted at 45° or −45°. All this makes
these diagrams very valuable. It is very difficult to think without them. With the Penrose diagrams, we see things much more
clearly than without them. For instance, we saw how the Penrose
diagram of figure 23 is much more talkative than the more conventional diagram of figure 24.

If you play with the Penrose diagrams, it won't take you long to
become familiar with them and to be able to use them efficiently.

You may say that the Penrose diagrams are abstract pictures. Yet
general relativity is still classical physics. It is much less abstract
than for instance quantum mechanics or quantum field theory.
General relativity can be represented. It is more or less easy of
course, but in the end usual intuition works. Not so in quantum
mechanics, which, as we saw in volume 2 of TTM, is much further
away from ordinary experience.

In the next lecture we shall finally arrive at Einstein's field equations, which are the keystone of general relativity.

Lev Landau[15] said of general relativity that it is the most beautiful
physical theory ever built.

---

[15]Lev Landau (1908–1968), Russian theoretical physicist. He coined the
expression "The Theoretical Minimum" – used as the name of Professor
Susskind's series of courses and books – to speak of what one should know
to start doing physics. When Lev Landau made his comment, he spoke, of
course, of the theories up to his time.

Lecture 9: Einstein Field Equations

Andy arrives, big earphones on his head, listening to music.

Lenny: *Relaxing before today's chat?*

Andy: *I'm in awe of the beauty of general relativity. I used to think that physics was a collection of nice theories to explain Newton's cradle or the lift of a wing wonderfully using complex variables or other stuff like that. But it's like comparing a popular song on the guitar and Bach's* St Matthew Passion. *So I'm listening to it.*

Lenny: *Wait until today's ninth lesson on Einstein field equations! It'll appropriately be like Beethoven's* Ninth Symphony.

Introduction
Newtonian gravitational field
Continuity equation
Energy-momentum tensor
Ricci tensor and curvature scalar
Einstein tensor and Einstein field equations
Questions/answers session

Introduction

We finally arrive at Einstein field equations. We will introduce the continuity equation, which says that certain things cannot disappear from here to reappear there without passing in between, and the energy-momentum tensor, which extends the idea of distribution of masses in space and their motions. Then we will be able to derive Einstein's equations, which are the analog in general relativity to Newton's equations of motion in classical non-relativistic physics, in their version using the concept of field.

We are not going to get deep into solving Einstein field equations. They are mathematically rather complicated. Even writing them down explicitly is intricate.

That is a feature of general relativity we have already mentioned: the principles are pretty simple, but the equations are computationally nasty. Almost everything we try to calculate gets complicated fast. There are lots of independent Christoffel symbols, elements of the curvature tensor, derivatives, etc. Each Christoffel symbol has a bunch of derivatives. The curvature tensor has more derivatives. The equations become hard to write on a single piece of paper.

The best way to solve them, or even write them, is just to feed them into your computer. And Mathematica[1] will spit out answers whenever it can. As said, the basic principles are simple, but going anywhere past the basic principles tends to be computationally intensive.

So we won't do much computation. We will concentrate on the meaning of the symbols. And then we will see what happens when we try to solve them in various circumstances.

In the final lecture, we will solve the equations in a simple case when we talk about gravitational waves. But in the present lecture, the topic is not gravitational waves, it is the fundamental equations of general relativity: Einstein field equations.

Newtonian Gravitational Field

Before we talk about Einstein field equations, we should talk first about the corresponding Newtonian concepts. Newton didn't think in terms of fields.[2] He didn't have a concept of field equations. Nevertheless, in classical non-relativistic physics, there are field equations that are equivalent to Newton's equations of motion.

---

[1]Symbolic mathematical computation program created by Stephen Wolfram (born 1959), English computer scientist.

[2]The invention of the concept of a field in physics is usually credited to the British physicist Michael Faraday (1791–1867).

Field equations of motion are always a sort of two-way street. Masses affect the gravitational field and the gravitational field affects the way masses move. Let's talk about this interplay in the context of Newton.

First of all, a field affects particles. That is just a statement that a gravitational force F can be written as minus a mass m times the gradient of the gravitational potential. The gravitational potential is usually denoted as ϕ, which is a function of position x. We write

$$F = -m \, \nabla\phi(x) \qquad (1)$$

This equation means that everywhere in space, due to whatever reason, there is a gravitational potential $\phi(x)$ that varies from place to place. You take the gradient of $\phi(x)$. You multiply it by m, the mass of the object whose motion you want to figure out, and tack on a minus sign. That tells you the force on the object.

Equation (1) can be one-dimensional, in which case the gradient is just the ordinary derivative of the function ϕ with respect to the independent space variable x. Then equation (1) just equates two scalars.

Equation (1) can also be multidimensional. It is the case when x is a vector position, which we could then denote $X = (x, \, y, \, z)$. The sign ∇, which reads "gradient" or "del," is used to express the vector of partial derivatives

$$\nabla\phi(X) = \left(\frac{\partial\phi}{\partial x}, \, \frac{\partial\phi}{\partial y}, \, \frac{\partial\phi}{\partial z} \right)$$

In that case, of course, the force F is also a vector. For simplicity, however, we will continue to use the notation x, be it one-dimensional or multidimensional. Same remark for F and for the acceleration that we shall presently introduce.

Equation (1) is one aspect of the field $\phi(x)$. It tells particles how to move. In the case that concerns us, it does so by telling them what their acceleration should be. This comes from Newton's equation linking the force F exerted on the particle to its acceleration a:

$$F = ma \qquad (2)$$

Combining equations (1) and (2), the mass m cancels. We obtain

$$a = -\nabla\phi(x) \qquad (3)$$

That is one direction of the two-way street. It is how a field tells a particle how to move.

On the other hand, that is, in the other direction, masses in space tell the gravitational field what to be. The equation that tells the gravitational field what to be is Poisson's equation.[3] It says that the second derivative of ϕ with respect to space (or the sum of second partial derivatives with respect to space in the multidimensional case) is related to the distribution of masses in space as follows:

$$\nabla^2\phi = 4\pi G \,\rho(x,\ t) \qquad (4)$$

Let's explain how to read equation (4):

(a) As said, if we are in several dimensions,[4]

$$\nabla^2\phi = \frac{\partial^2\phi}{\partial x^2} + \frac{\partial^2\phi}{\partial y^2} + \frac{\partial^2\phi}{\partial z^2}$$

(b) The multiplicative factor 4π is a convention. It originates from the geometry of the sphere, because we often deal with a field that is invariant under rotation, i.e., spherical.

(c) G is Newton's gravitational constant. It is equal to

$$6.674 \,\times\, 10^{-11} \text{ N m}^2/\text{kg}^2$$

(d) Finally, ρ is the *density of mass* at each position x, or $(x,\ y,\ z)$. It can depend also on time because masses can move around.

Equation (4) expresses how the distribution of masses in space determines the gravitational field.

---

[3] It is one of the great equations of mathematical physics, named after Siméon Denis Poisson (1781–1840), French mathematician and physicist.

[4] ∇^2, which is a notation for the formal dot product $\nabla \cdot \nabla$, is called the *Laplacian operator* and has its own notation: it is sometimes denoted $\triangle$.

We just saw the two aspects: the field tells particles how to move, and masses, particles in other words, tell the field how to be – how to curve, as we will see.

In general, the density ρ, which is by definition the amount of mass per unit volume, will be a complicated function over the whole space, depending also on time, leading to complicated calculations. However, we can easily solve equation (4) in the special case when the distribution of masses (or the density[5]) ρ consists simply in all the mass concentrated at one point.

In the case when $\rho(x)$ is "concentrated" like that, it is actually a Dirac distribution[6] – we met these extended or generalized mathematical functions in volume 2 of TTM. They are manipulated mostly through their integral or the integral of their product with some other function over some region.

At first, we may be considering a real star or a planet or a bowling ball of mass M, as in figure 1. The only constraint is that it must be a spherically symmetric object of total mass M. It does not matter whether the mass is uniformly distributed inside the volume provided it is symmetric with respect to rotation. Like in an onion, there could be layers with different density, for instance more density near the center than near the outer surface.

Figure 1: Spherically symmetric object of mass M, equivalent, according to Newton's theorem, to a point mass of same mass M concentrated at the center if we are outside the object.

---

[5]The mass in any small volume around a point x being then the local density of mass $\rho(x)$ multiplied by the small volume of space dx or $dx\,dy\,dz$.

[6]Named after Paul Dirac (1902–1984), British theoretical physicist and mathematician, who used them first.

Anyhow, once we are outside the region where the mass is, according to Newton's theorem, which we saw in the previous lecture, we can treat the body as a point mass, that is, a Dirac density (another name for a Dirac distribution) with its peak at the center.

The solution to equation (4) in this case is easy. It is

$$\phi(r) = -\frac{MG}{r} \qquad (5)$$

where we replaced the spatial coordinate x by a radius r.

If we consider that the body is a point mass, then equation (5) is the valid solution to equation (4) everywhere except at the central point (where ϕ is not defined since $-MG/r$ tends to infinity).

Let's explore further this example where ϕ is given by equation (5). Taking its gradient just puts another r downstairs, while changing the sign. Indeed, differentiating $1/r$ with respect to the radius, which is what ∇ does, produces $-1/r^2$. We wind up with

$$F = \frac{mMG}{r^2} \qquad (6)$$

This is a kind of primitive field theoretic way to think about gravitation. *Instead of action at a distance – a concept that modern physics rejects – we have a gravitational field.*

In truth here our field still implies action at a distance, because in Newton's theory if you move a mass around, ϕ instantly reacts to it and changes. This is seen in equation (4). But it is a way of writing the theory, making it look like a field theory.

Equations (3) and (4) correspond respectively to these relations:

- equation (3): field $\rightarrow$ motion of mass,

- equation (4): mass $\rightarrow$ structure of field.

It is these equations we want to replace by something making sense in general relativity.

To begin to understand what we are going to do, let's remember the Schwarzschild geometry. In lecture 5, we pulled the Schwarzschild geometry out of a hat – it was a given. Of course, it was the solution to Einstein's equations in the case of a central point mass field. But we did not have the tools to establish it yet.

Equations (3) and (4) are not Einstein's equations, they are Newton's equations. Their solution, in the simple case of a central spherically symmetric field, is $\phi(r) = -MG/r$.

In lecture 5, we wrote down the Schwarzschild metric, see equation (32) of that lecture. We are not going to write the whole shebang again. But let's examine what g_{00} was:

$$g_{00} = 1 - \frac{2MG}{r} \qquad (7)$$

As usual, we have set the speed of light c equal to 1.

Our aim presently is to guess some correspondence between g_{00}, given by equation (7), and anything in the Newtonian framework.

Using equation (5), the above equation (7) can be rewritten as $g_{00} = 1 + 2\phi$. Therefore $\nabla^2 g_{00}$ is just twice $\nabla^2\phi$. This leads to

$$\nabla^2 g_{00} = 8\pi G\rho \qquad (8)$$

This formula should be taken with a grain of salt. It is just a mnemonic device to remember the relationship between some aspects of general relativity and matter.

Nonetheless, interestingly enough, it already suggests that matter or mass is affecting geometry. When we make this correspondence between Newton's ϕ and the Schwarzschild metric, we see roughly that matter is telling geometry how to curve, so to speak. Roughly only, because we are going to be more precise.

Equation (8) is not really the part "mass → structure of field" of Einstein's equations. It is a good deal more complicated, but we begin to get a sense of where we are heading.

What about the other part "field → motion of mass"? In Newton's framework, we saw equation (1) or (2), which specifies how the field dictates the motion of mass. In general relativity, the equation $\ddot{x} = - \nabla\phi(x)$ is replaced by the statement that once we know the geometry, i.e., once we know g_{00}, the rule is:

Particles move on space-time geodesics.

For the other direction, we saw that the Newtonian field equation,

$$\nabla^2 \phi = 4\pi G\rho$$

is replaced by something we naively wrote as

$$\nabla^2 g_{00} = 8\pi G\rho$$

We are going to do better. We have to figure out exactly how the mass distribution affects the field. A little more than a century ago, Einstein figured it out exactly.

Before we write down the field equations, we need to understand more about the right-hand side of each equation: $4\pi G\rho$ in Newton field equations, or $8\pi G\rho$ in Einstein field equations. We need to develop our knowledge of the density of mass.

Mass really means energy. Indeed, we saw in volume 3 on special relativity that $E = mc^2$. If we forget about c – that is, if we set it equal to one – then energy and mass are the same thing. Therefore what goes on the right-hand side of the field equations is really energy density.

The question is: what kind of quantity in relativity is the energy density? It is part of a complex of things that includes more than just the energy density. It is one component of some tensor, the other components of which have other meanings.

Let's go back and review quickly the notion of conservation in physics. In the present case, it will be conservation of energy and momentum. But we will begin with a simpler case: conservation of charge.

Continuity Equation

Conservation, densities, flows of things like charge and mass, etc. – let's review briefly these concepts.

To begin with, consider the electric charge. It is simpler than energy for reasons we will come to.

The total electric charge of a system is called Q. It is the standard notation for the electric charge. In many situations the electric charge *density* is called ρ, but to avoid confusion with the mass density or the energy density, which we already call ρ, we shall denote the charge density with the letter σ.

What is electric charge density? Consider a small volume of space, a differential volume of space at a given point. Take the electric charge in that volume, and divide it by the volume. This gives you the electric charge density at that point. The density can be schematically written

$$\sigma = \frac{Q}{\text{volume}} \tag{9}$$

It has units of charge divided by volume.

Let's look at an infinitesimal volume, as in figure 2.

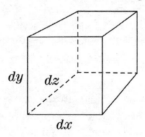

Figure 2: Infinitesimal volume in 3D space.

If we look at electric charges moving around in space, between time t_1 and time t_2, we see some entering the volume $dx\ dy\ dz$, some leaving, some passing by, some passing through, etc. This is shown in figure 3.

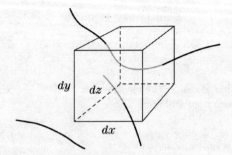

Figure 3: Trajectory in space, between time t_1 and time t_2, of various moving electric charges.

The current in direction x is defined as the quantity of charge passing through the side $dy\,dz$ per unit of time. It is denoted J^x. There are analogous definitions and notations for y and z.

Similarly to equation (9), notice that each current along one spatial dimension can be schematically written

$$J^m = \frac{Q}{\text{area} \times \text{time}} \tag{10}$$

In relativity, however, we have four dimensions more or less on an equal footing. Time and space are nice and symmetric to each other. So dividing by area times time is again dividing by three dimensions. One happens to be time-like, two happen to be space-like.

Both equations (9) and (10) can be thought of as a charge divided by a three-dimensional volume. In the case of equation (9), it is a pure spatial volume. In the case of equation (10), it is a mixed space-and-time three-dimensional thing.

The combined density σ plus the three currents J^x, J^y, and J^z happen to form a 4-vector in the sense of relativity.[7] Following the standard notation we became familiar with in volume 3, instead

---

[7] In a change of coordinates, they change according to the tensor equations of a change of frame of reference, which we studied in lecture 2.

of $(\sigma,\ J^x,\ J^y,\ J^z)$ we simply write

$$J^\mu \qquad (11)$$

where the Greek index μ runs from 0 to 3. J^0 (read "J naught") is the electric charge density σ. And the other three elements are the components of the current indexed with 1, 2, and 3 instead of x, y, and z.

We arrive at an important physical fact:

The conservation of electric charge is a local property.

We haven't yet talked about the familiar relationship between the evolution of the charge Q inside the volume $dx\ dy\ dz$ and the current going through its sides, because we want to stress that, in theory, conservation could hold in a system without being local. The electric charge carried by an object in a system could at time t disappear from the object and instantaneously reappear elsewhere in a distant part of the system. If the system we are considering is the entire universe, the charge could disappear from my desk and reappear in Alpha Centauri. I always use Alpha Centauri as someplace so far away that it doesn't matter.

If that were possible, conservation would still hold. You would say rightly: "Well, charge is conserved." I would retort: "Who cares if charge is conserved, if it can just disappear to some arbitrarily distant place. It is just as good as saying it wasn't conserved."

In the laboratory, however, charge doesn't disappear that way. If it leaves the laboratory, it goes through the walls or the windows or the roof, or simply through the door. In other words, the electric charge cannot change in a given volume without some current flowing through the boundary.

That idea is called *continuity*. There is an equation that goes with it, called the *continuity equation*.

If we look at a box of volume one in some units, the charge inside the box is σ, that is, the charge density times one. The charge leaving the box over a unit time is $-\dot\sigma$. Thus it is minus the time derivative of σ. Why minus? Because it is leaving the box.

That has to be equal to the sum of the currents passing through the box. A basic theorem of multivariate calculus, called the *divergence theorem*, also known as Gauss's theorem or Ostrogradsky's theorem,[8] states that the diminution of charge inside the box is equal to the divergence of the current:

$$-\dot{\sigma} = \nabla \cdot J \tag{12}$$

The "formal dot product" $\nabla \cdot J$ on the right-hand side is a convenient way to mean $\frac{\partial J^x}{\partial x} + \frac{\partial J^y}{\partial y} + \frac{\partial J^z}{\partial z}$.

Remembering that t is one of the four components of space-time in relativity, this can be rewritten nicely, first as

$$\frac{\partial \sigma}{\partial t} + \frac{\partial J^x}{\partial x} + \frac{\partial J^y}{\partial y} + \frac{\partial J^z}{\partial z} = 0 \tag{13}$$

Then, since J^0 is by definition σ, see expression (11), using the summation convention we can write this even more nicely as

$$\frac{\partial J^\mu}{\partial X^\mu} = 0 \tag{14}$$

The divergence theorem has led us to a simple tensor-type equation satisfied by J^μ.

J^μ is a 4-vector. The X^μ's are the four components of space-time. Therefore equation (14) has the earnest look of a good equation involving the derivative of a tensor with respect to position. Since equation (14) is true in any frame (because it is a tensor equation), it expresses a conservation law.

Note that in curved coordinates we would have to be more careful. Equation (14) would be correct only in ordinary flat coordinates. In curved coordinates, we might have to replace the ordinary derivative by the *covariant derivative* of the tensor:

---

[8]It was actually first stated by Lagrange in 1762, then independently by Gauss in 1813, Ostrogradsky in 1826, Green in 1828, etc. It is the multivariate generalization of $\int_a^b f(x)dx = F(b) - F(a)$.

$$\frac{DJ^\mu}{DX^\mu} = 0 \tag{15}$$

Remember the covariant derivative of tensors, which we studied in lecture 3. It turns out that in the case of electric current it doesn't matter. But in general it would matter.

So when we go to curved coordinates, we should replace all ordinary derivatives by covariant derivatives, otherwise the equations we get are not good tensor equations.

Why do we want tensor equations? We want such equations because we want them to be true in any set of coordinates.

That is the theory of electric charge, flow, current, and the continuity equation. Equation (15) is called the continuity equation for electric charge. The physics of it is that when charge either appears into or disappears from a volume, it is always traceable to a current flowing in or out through the boundaries of that volume.

Let's now turn to energy and momentum, which are a bit more involved.

Energy-Momentum Tensor

Energy and momentum are also conserved quantities, though not as simply as electric charge. We will see that energy and momentum altogether form a 4-vector.

Like charge, they can be described in terms of density of energy and density of momentum, that is, density of each component of momentum. We can ask how much energy there is in a volume, in the form of particles or whatever it happens to be, including the mc^2 part of energy. We can ask how much momentum is in a volume. Just look at all the particles within a volume and count their momentum.

Photons, or electromagnetic radiations, have both energy and momentum. That energy and momentum can be regarded as the integral of a density. So in that sense, each one of them, each energy and each component of the momentum, is like the electric charge. They are conserved. They can flow. If an object is moving, the energy and momentum will be flowing too.

The question is: how do we represent the same set of ideas for energy and momentum as we just did for electric charge?

Now there is a difference between, on the one hand, charge and, on the other hand, energy and momentum. Let's talk first about charge. Electric charge is an invariant. No matter how the charge is moving, the charge of an electron is always the same. It does not depend on its state of motion.

Pursuing on charge, the density of charge and the current of charge, however, are not invariants. For example, suppose we have a given charge, taking a certain volume of space, and I decide to look at it in a different frame of reference than you. I run by it with a certain velocity. Because of Lorentz contraction I say that the volume of that charge is one thing. You, sitting still relative to the charge, assign a different volume to it. Since the charge itself does not change, we will not agree about the value of the charge density. The currents will also obviously be different. Indeed, in my frame of reference I can see some charges moving, while you see them at rest. You say those charges create no current. I'm moving and I see a wind of charges passing me by. So I say there is current. We are both right of course because charge density and charge current are not invariant.

The solution was to observe that charge density, J^0, and the three dimensions of current, J^n, are the components of a 4-vector (see volume 3 of TTM to refresh your memory about 4-vectors). Depending on the reference frame, their components change according to tensor equations of change of frame. But if the 4-vector is zero in one frame, it is zero in every frame.

Let's turn to energy and momentum. The situation is a bit more complicated. The total energy and momentum – I'm not talking now about the density of them but about the total energy and momentum – are not invariant.

I see a particle standing still – I'm talking about the whole particle
not the density. I say there is some energy of a certain magnitude.
You are walking past it and you see not just the $E = mc^2$ part of
the energy, but you also see kinetic energy of motion.

You see more energy in the particle or object not because of any
Lorentz contraction of the volume it is in, but because the same
object when you look at it has more energy than when I look at it.

The same is true of the total momentum, not the flow of it, not
the density of it, but the momentum itself. It is also frame dependent. You see an object in motion, you say there is momentum
there. I see the same object at rest, I say there is no momentum.

So energy and momentum, unlike charge, are not invariant. They
together form the components of a 4-vector:

$$P^\mu = (E, \ P^m) \tag{16}$$

The quantity E is the energy, also called, naturally enough, P^0.
And the P^m are the components of momentum, where the Roman
index m labels the directions of space. So zero is the index for
energy, and the others are for momentum.

Each of the four components of P^μ is like a charge Q. In other
words, it is a conserved quantity in the sense of invariance of 4-
vectors. Let's repeat the tensor rule we are referring to: if two
4-vectors are equal in one frame, they are equal in every frame.

Remember, in equation (14) or (15), density was the zeroth component of J^μ. So zero is also the index for density, and the others
are for flow.

Let's turn to the energy in P^μ. As we just said, it is P^0. Consider
now the *density of energy*. In other words, how much energy is in
a small volume. We are going to denote this as follows:

$$T^{00} \tag{17}$$

The first index 0 refers to energy as opposed to momentum, and
the second 0 to the fact that we are considering a density.

It is a function of position, so we will sometimes write explicitly

$$T^{00}(X) \tag{18}$$

where X stands for $(X^0,\ X^1,\ X^2,\ X^3)$.

Energy can also move. Like charge it can disappear from a region, and like charge too it does so by crossing the boundary of the region. In other words, the continuity equation applies to energy density and energy flow, as it applied to charge density and electric current.

The amount of energy flowing through a unit surface, along the X^1 direction, per unit time – the "current of energy" you might call it – is denoted

$$T^{01}(X) \tag{19}$$

The first index, that is, index 0, refers to an energy, and the second index, that is, index 1, to a flow along the first spatial axis. Likewise there is a T^{02} and a T^{03}.

In summary, the three components $(T^{01},\ T^{02},\ T^{03})$ form the flow of energy, while T^{00} is the density of energy.

The continuity equation for energy is derived in exactly the same way as we did for charge. It involves now the components $T^{0\nu}$. And what does it say? For the moment the first index 0 is just passive. It signals that we are talking about energy. It is the second index ν running over 0, 1, 2, and 3, which is like the index μ of J^μ in equation (15). Since ν is a dummy index, we may as well use the letter μ.[9] The *continuity equation for energy* then is

$$\frac{DT^{0\mu}}{DX^\mu} = 0 \tag{20}$$

Next step: everything that we said about energy we could now say about any one of the components of the momentum.

---

[9] Go back to the mathematical interlude on dummy variables in lecture 1, if you are unsure about what they are.

Let's go to the momentum components of the 4-vector P^μ. These are the last three components, customarily indexed with the Roman letter m running from 1 to 3.

The component P^m also has a density, denoted T^{m0}. The naught indicates that it is a density. The m indicates that we are talking about the m-th component of momentum. It is also a function of the position X in space-time.

Likewise we can consider the flow of the m-th component of momentum, along the n-th direction of X. It is denoted T^{mn}.

Momentum is a conserved quantity, in the sense that, together with the energy, the three momenta form a 4-vector. Momentum can flow. Each of its components can flow along some direction.

We can then go a little further. We can say that the same equation as equation (20) is true even if we replace energy by a component of momentum. In other words we could replace naught by n. Now we have the same equation for all four possibilities of the first index. So, switching μ and ν again for cosmetic reasons, we can finally write

$$\frac{DT^{\mu\nu}}{DX^\nu} = 0 \qquad (21)$$

Let's step back and see where we arrived. The densities and flows of energy and momentum form a tensor with two indices. The first index tells us whether we are talking about energy or momentum, the other index whether we are talking about density or flow. The matrix we obtain is called the *energy-momentum tensor*.

This energy-momentum tensor has an interesting property, which we have not proved. Take for example T^{m0}, that is, the density of the m-th component of momentum. Compare it with T^{0m}, that is, the flow of energy in the m-th direction. It is a general property of relativistic systems, which we are not going to prove, that the matrix formed by the components of the energy-momentum tensor is symmetric. In other words,

$$T^{m0} = T^{0m} \quad \text{and more generally} \quad T^{\mu\nu} = T^{\nu\mu}$$

Relativistic invariance allows you to connect T^{m0} with T^{0m}. It is a theorem of all relativistic field theories, which requires some work to prove: the energy-momentum tensor is symmetric. Here, let's take it for granted and write the energy-momentum tensor in matrix form:

$$T^{\mu\nu} = \left(\begin{array}{cccc} T^{00} & T^{01} & T^{02} & T^{03} \\ T^{01} & T^{11} & T^{12} & T^{13} \\ T^{02} & T^{12} & T^{22} & T^{23} \\ T^{03} & T^{13} & T^{23} & T^{33} \end{array} \right) \qquad (22)$$

We will come back to the meaning of these elements in a moment. The term T^{00} is clear: it is energy density. The terms (T^{01}, T^{02}, T^{03}) are fairly clear too: they are flows of energy. The term T^{10} (which is the same as T^{01}) is momentum density. Then we have flows of momentum. So their meaning is pretty clear. But we will find out that some of the elements of $T^{\mu\nu}$ have other meanings, connected with pressure and other notions of that nature.

The important idea is that *the flow and density of energy and momentum form an energy-momentum tensor*. And this energy-momentum tensor satisfies four continuity equations, one for each type of physical quantity we are talking about (energy, that is, P^0, or momentum along one of the three spatial axes, that is, P^m).

What we have learned is that the notion of energy density, ρ, which we tried to relate to the metric in equation (8), reproduced here,

$$\nabla^2 g_{00} = 8\pi G\rho \qquad (23)$$

is incomplete. It is part of a complex of things. It is part of a tensor, of course.

This is a fundamental observation: the right-hand side of equation (23) is part of a tensor, because ρ is T^{00}. So the left-hand side must also be part of a tensor.

In physics an equation which says that some particular component of a tensor is equal to the same component of some other tensor is usually meaningless or wrong – unless its full version says that the two tensors are equal. We already met this idea several times in lecture 1 and lecture 2, and in the other volumes of TTM. Because it plays an important role in the present lecture, to build up Einstein field equations, let's repeat it. It is easier to explain it with vectors, which are a simple type of tensor.

Suppose we have two 3D spatial vectors A and B (what we sometimes call 3-vectors). Suppose we assert that there is a law saying that A_3 is equal to B_3, the index 3 being the z direction. Does this make sense as a law of physics?

Well, it only makes sense as a law of physics if it is also true that $A_2 = B_2$ and $A_1 = B_1$. Why is that? Because if it is a law of physics, it means that it is true in every reference frame. But we can always rotate the coordinates so that the first axis in the old frame becomes the third axis in the new frame, so A_1 must also be equal to B_1, etc.

This is an illustration of the general idea that vectors and tensors are entities, geometric or otherwise, which exist independently of any consideration of coordinates or components. "Go five steps in this direction," if I show you a direction while telling you that, is meaningful without referring to coordinates. Of course, with appropriate coordinates, it may become "go three steps to the east, and four steps to the south." But these coordinates are frame-dependent. My injunction may have its first coordinate equal to 3 in one frame, but only equal to 1 in another.

When we go to relativity, the same considerations are true for 4-vectors, including the time component, because it can be transformed into a space component through a Lorentz transformation. They are also true for any two tensors of the same type.

To summarize: tensor equations, to be good laws of physics, must equate *all the components* of two tensors, not just two components of them. If we have an equation, which we believe should reflect a law of physics, that equates two components of tensors, then it should be true for all the components of the two tensors.

Equation (23) is a formula that equates the naught naught component of the energy-momentum tensor with something else. If it is to be a law of physics, we should figure out which tensors it actually equates.

Let's not worry too much about whether the left-hand side is correct or not. We just made a guess of what the left-hand side might look like, and found something related to the metric, that is, to the geometry.

The right-hand side is the energy density, and it is what you would expect to be on the right-hand side of *Newton's equations*.

Therefore the right-hand side of *Einstein's equations* must involve not a particular component of a tensor; it must generalize to something that involves all components of a tensor. That means Einstein's generalization of Newton must be as follows. The right-hand side should be

$$\ldots = 8\pi G T^{\mu\nu}$$

A special case of it is when both μ and ν are time. Then it becomes the energy density. But if the equation that we are constructing is to be true in every frame, it has to be a tensor equation involving all the μ ν components.

What must there be on the left-hand side? There must also be a rank-2 tensor. Otherwise the equation would not make sense. It must be symmetric because the right-hand side is symmetric. It must have whatever other properties the right-hand side has.

But the left-hand side will not be something made up out of matter. It will be made up out of the metric. *It will have to do with geometry, not with masses and sources.*

We shall call the left-hand side $G^{\mu\nu}$, and we shall pursue our guess work on the equation

$$G^{\mu\nu} = 8\pi G T^{\mu\nu} \tag{24}$$

The only thing we know about $G^{\mu\nu}$ is that it is made up out of the metric. It probably has second derivatives in it to be the analog of the Laplacian of Newton's equation (4). In other words, it will involve the metric in some form or other and very likely second derivatives of the metric.

Ricci Tensor and Curvature Scalar

We began to see what kind of object we would like to have on the left-hand side of equation (24), which will generalize Newton's field equation in the general relativity setting of Einstein.

When we find a good candidate for it, in order to evaluate it we can ask the following question: when we are in a situation where non-relativistic physics should be a good approximation, does this $G^{\mu\nu}$ reduce to just $\nabla^2 g_{00}$? Perhaps it does. Perhaps it doesn't. If it doesn't, then we throw it away and try to find a different candidate.

Let's explore the possibilities for $G^{\mu\nu}$. It is a tensor made up out of the metric. It has second derivatives, or at least it must have some terms that have second derivatives. So it is not the metric by itself. What kind of tensor can we make out of a metric and second derivatives? We already met one: the curvature tensor.

Recall what is the curvature tensor. It is made up out of the Christoffel symbols, which themselves are functions of the metric.

Let's start with the Christoffel symbols

$$\Gamma^{\sigma}_{\nu\tau} = \frac{1}{2} \, g^{\sigma\delta} \left[\, \partial_{\tau} g_{\delta\nu} + \partial_{\nu} g_{\delta\tau} - \partial_{\delta} g_{\nu\tau} \, \right] \qquad (25)$$

The only important thing in this equation is that the right-hand side involves the first derivatives of g terms.

Next, let's write the curvature tensor in all its glory.[10]

$$\mathcal{R}^{\sigma}_{\mu\nu\tau} = \partial_{\nu}\Gamma^{\sigma}_{\mu\tau} - \partial_{\mu}\Gamma^{\sigma}_{\nu\tau} + \Gamma^{\lambda}_{\mu\tau}\Gamma^{\sigma}_{\lambda\nu} - \Gamma^{\lambda}_{\nu\tau}\Gamma^{\sigma}_{\lambda\mu} \qquad (26)$$

---

[10]It is the same as expression (25) in lecture 3, with different dummy indices.

This rank-4 tensor is the object – the "mathematical probe" –
that tells us whether there is real curvature in space-time, not
just curvature due to curvilinear coordinates that could be flattened (like a uniform acceleration can be flattened; see lecture 1,
figure 4, and lecture 4, figure 15). If any component of the curvature tensor is nonzero at any point or in any region of space, then
the space is curved.

The right-hand side of equation (26) uses the summation convention for the index λ in the last two terms. Again, the only
important point in the expression of $\mathcal{R}_{\mu\nu\tau}{}^{\sigma}$ is that it involves another differentiation.

The Christoffel symbol – which, the reader remembers, is not a
tensor – involves first derivatives of the metric. And the first two
terms of the curvature tensor are first derivatives of the Christoffel
symbols. Therefore the curvature tensor has second derivatives of
the metric in its first two terms, and squares of first derivatives in
its last two terms.

Therefore $\mathcal{R}$ is a candidate, or various functions of it are candidates, to appear on the left-hand side of equation (24). But wait!
The Riemann curvature tensor $\mathcal{R}_{\mu\nu\tau}{}^{\sigma}$ has four indices, whereas the
energy-momentum tensor has only two.

What can we do with the curvature tensor to transform it into
a thing with only two indices? We can contract it.[11] Remember
the rule: if we set the upper index σ equal to one of the lower
indices and apply the summation convention, we eliminate two
indices and get a rank-2 tensor. If we used τ for the lower index
in the contraction, we would discover that we get zero. The vari-
ous symmetries and minus signs in the expression of $\mathcal{R}_{\mu\nu\tau}{}^{\sigma}$ would
wind up yielding 0.

However if we set σ and ν equal, we won't get zero. We will get
something that we can call the tensor $\mathcal{R}_{\mu\tau}$. So from a tensor with
four indices, by contraction we can build a tensor with two indices.
But we have to be careful not to get zero.

---

[11] We saw in lecture 2 that, for instance, the dot product of two vectors V
and W is the contraction of the tensor $V^m W_n$. And this can be generalized.

We won't get the zero tensor when we contract σ with ν or σ with μ. In fact the two tensors we get by contracting σ with ν and σ with μ happen to be the same tensor except for the sign.

So there is only one tensor that we can build out of two derivatives acting on the metric and that has only two indices. It is actually a well-known theorem.

This tensor is called the *Ricci tensor*, in honor of Gregorio Ricci-Curbastro whom we mentioned in lecture 1. It is a contraction of the Riemann tensor. The Riemann tensor has a lot more components. The Ricci tensor carries less information. As a consequence, it can be zero while the Riemann tensor is not zero.

As always, if you have a tensor, you can raise and lower its indices. That means there are also things denoted $\mathcal{R}^{\mu}_{\tau}$, $\mathcal{R}^{\tau}_{\mu}$, and $\mathcal{R}^{\mu\tau}$. Recall that we raise and lower indices using the metric tensor.

Another fact about the Ricci tensor is that it happens to be symmetric. And this is true of its version with upper indices as well:

$$\mathcal{R}^{\mu\tau} = \mathcal{R}^{\tau\mu} \tag{27}$$

This can be checked from its definition. I don't know any simple quick argument. All these tensors have fairly complicated expressions, but checking the facts we stated is most of the time pretty mechanical.

The Ricci tensor $\mathcal{R}^{\mu\tau}$, we said, is symmetric. The right-hand side of equation (24) is a rank-2 tensor, with two contravariant indices, and is symmetric too. So the Ricci tensor $\mathcal{R}^{\mu\tau}$ is a candidate for the left-hand side:

$$\mathcal{R}^{\mu\nu} = ? \; 8\pi G T^{\mu\nu} \tag{28}$$

There is another tensor that we can make out of the Ricci tensor. It is by contracting its two indices (after having raised or lowered one of them). We define

$$\mathcal{R} = \mathcal{R}^{\mu}_{\mu} = \mathcal{R}^{\mu\tau} g_{\mu\tau} \tag{29}$$

The quantity $\mathcal{R}$ defined by equation (29) is called the *curvature scalar*. It is a scalar tensor. It has no indices left. So it's not something we want on the left-hand side of equation (24). But we can multiply it by something suitable.

For instance, we can multiply it like this: $g^{\mu\nu}\mathcal{R}$. This does give us again a tensor of the type we are looking for. So another possibility would be

$$g^{\mu\nu}\mathcal{R} \; = \; ? \;\; 8\pi G T^{\mu\nu} \tag{30}$$

I'm not recommending either of these at the moment. I'm just saying, from what we have said so far, that either of the formulas we reached could be possible laws of gravitation. The left-hand sides of equation (28) and (30) both involve second derivatives of the metric tensor, equated to something on the right-hand side that looks like a density and flow of energy and momentum.

Which one should we pick? Well, we know one more thing. It is the conservation of energy and momentum or, better yet, the continuity equation for energy-momentum. If we believe that energy and momentum have the property that they can only disappear by flowing through walls of systems, then we are forced to conclude that

$$D_\mu T^{\mu\nu} = 0$$

That is the continuity equation. It follows that we must also have

$$D_\mu G^{\mu\nu} = 0 \tag{31}$$

The first thing we could do is check whether either of the candidate solutions to equation (24), that is, those in equation (28) or equation (30) satisfy condition (31). If not then the left-hand side simply cannot equal the right-hand side, unless we give up local continuity of the energy and momentum.

Let's check the left-hand side of equation (30). We want to calculate

$$D_\mu \; (g^{\mu\nu}\mathcal{R})$$

Covariant derivatives satisfy the usual product rule of differentiation. So we have

$$D_\mu \; (g^{\mu\nu}\mathcal{R}) = (D_\mu \; g^{\mu\nu})\mathcal{R} + g^{\mu\nu}(D_\mu \; \mathcal{R}) \tag{32}$$

First of all, the covariant derivative of the metric tensor is zero. It is a consequence of the definition of covariant derivative; see lecture 3. Covariant derivatives are by definition tensors that in the special good frame of reference are equal to ordinary derivatives.

The good frame of reference is by definition the frame of reference in which the derivative of g is 0. So the first term on the right-hand side of equation (32) is eliminated.

Secondly, $\mathcal{R}$ is a scalar. The covariant derivative of a scalar is just the ordinary derivative. So we can rewrite

$$D_\mu \left(g^{\mu\nu} \mathcal{R} \right) = g^{\mu\nu} \, \partial_\mu \mathcal{R} \tag{33}$$

Certainly, in general, the derivative of the curvature is not zero. We know that there are geometries that are more curved in some places and less curved in others. So it cannot be the case that $\partial_\mu \mathcal{R}$ is identically equal to zero. The factor $g^{\mu\nu}$ doesn't change anything. Therefore we are led to eliminate the candidate $g^{\mu\nu}\mathcal{R}$.

What about the other candidate, the Ricci tensor with upper indices?

We do the same thing. We calculate

$$D_\mu \, \mathcal{R}^{\mu\nu}$$

The calculation is a little harder, but not much. The answer is

$$D_\mu \, \mathcal{R}^{\mu\nu} = \frac{1}{2} \, g^{\mu\nu} \, \partial_\mu \mathcal{R} \tag{34}$$

Again, it cannot be zero, for the same reason that the right-hand side of equation (33) can't be zero. The covariant derivative of the Ricci tensor happens to be exactly one half that of the other candidate, $g^{\mu\nu}\mathcal{R}$.

But now, from equations (33) and (34), we know the answer!

Einstein Tensor and Einstein Field Equations

Now we know which tensor on the left-hand side of equation (24) will satisfy all the conditions, including having covariant derivative zero. The left-hand side of the gravity equation (24) that will work is

$$G^{\mu\nu} = \mathcal{R}^{\mu\nu} - \frac{1}{2}\, g^{\mu\nu}\, \mathcal{R} \qquad (35)$$

A theorem can be proved that says there is no other tensor (up to a multiplicative factor) made up out of two derivatives acting on the metric that is covariantly conserved.

Of course, we could have twice $G^{\mu\nu}$ or half of it, or seventeen times it. But now it just becomes a question of matching equation (24), with the $G^{\mu\nu}$ we found, to Newton's equations in the appropriate approximation, that is, where everything is moving non-relativistically. Either there is some correct numerical multiple that will ensure the match or there isn't. If there isn't, then we're in trouble.

When we say that we want to match formulas, we mean that we take equation (24), we look at the time-time component of $G^{\mu\nu}$ given by equation (35), and we want, with some appropriate multiplicative factor in front of $G^{\mu\nu}$, in the non-relativistic limit – that is, everybody moving slowly, not too strong a gravitational field – to find Newton's equation (4).

The answer is that it works. And the multiplicative factor happens to be simply 1. It is just a piece of luck that it is not some other number. There is nothing deep about this factor.

We finally arrived at the equation we were looking for, generalizing Newton's equation:

$$\mathcal{R}^{\mu\nu} - \frac{1}{2}\, g^{\mu\nu}\, \mathcal{R} = 8\pi G\, T^{\mu\nu} \qquad (36)$$

The path that we followed in this lecture is essentially Einstein's calculations. He knew pretty much what was going on. But he didn't quite know what the right equation linking the energy-momentum tensor and the geometry was. I believe in the beginning he actually did try $\mathcal{R}^{\mu\nu} = T^{\mu\nu}$. He eventually realized that it didn't work. And he searched hard for a suitable left-hand side. I don't know how many weeks of work it took him to do all of this,[12] but in the end he discovered $G^{\mu\nu}$.

$G^{\mu\nu}$ is the *Einstein tensor*, $\mathcal{R}^{\mu\nu}$ is the *Ricci tensor*, $\mathcal{R}$ is the *curvature scalar*. Equation (36) is now known as the *Einstein field equations*. They generalize Newton's equations. And in the appropriate limit, they do reduce to Newton's equations.

The continuity equation has played a fundamental role in this derivation. It is what has led us to look only for candidates, on the left-hand side, that are covariantly conserved. Notice that the principle is as fundamental in classical non-relativistic physics as in relativity. It says that conservation of energy or of momentum is a local property. Things cannot disappear here and immediately materialize in Alpha Centauri. They have to travel through intermediate locations.

There is another way to derive Einstein field equations, using the action principle. It is much more beautiful and much more condensed. We introduce the principle of least action for the gravitational field. The calculations are harder than what we did. But in the end the field equations just pop out.

Equation (36) reveals something interesting. We see that in general the source of the gravitational field is not just energy density. It can involve energy flow. It can involve momentum density, and it can even involve momentum flow.

---

[12]In early summer 1915, at the invitation of Hilbert, Einstein gave a series of lectures in Göttingen, exposing his work of the past few years and his present problem to link sources and geometry. This led, during the summer and fall of 1915, to an exchange of friendly letters, but also to a race, between Einstein and Hilbert. Einstein published his final correct equations on November 25. Simultaneously Hilbert attained the same equations via the Lagrangian approach (see lecture 10). That is why the name of Hilbert is sometimes attached next to Einstein's to the field equations of general relativity. However Hilbert never denied that all the credit for general relativity should go to Einstein.

As a rule, however, the momentum components, and even the energy flow, but certainly the momentum flow and the momentum density, are much smaller than the energy density. Why can we say that? It has to do with the impact of the speed of light in the formulas. When we put back the speed of light (that is, when we use ordinary SI units), the energy density is always huge because it gets a factor c^2, like in $E = mc^2$.

On the other hand, momentum is typically not huge because it is just mass times velocity. When you are in a non-relativistic situation, when velocity is slow, energy density is by far the biggest component. The other components of the energy-momentum tensor are much smaller. They are typically smaller by powers of v/c.

In other words, in a frame of reference where the sources of a gravitational field are moving slowly, the only important element on the right-hand side of equation (36) is T^{00}, that is, the energy density ρ.

It is also true that, in the same limit, the only important element on the left-hand side is the second derivative of G^{00}. So in the non-relativistic limit these elements match and reduce to Newton's equations.

If you are outside the non-relativistic limit, however, in places where sources are moving rapidly, or even places where the sources are made up of particles that are moving rapidly, even though the whole system may not be moving so much, other components of the energy-momentum tensor do generate gravitation.

Since we now know that, in a sense, gravitation is just geometry, we see that in relativistic situations all the components of the energy-momentum tensor do participate in the generation of the curvature. It is not just energy – or mass – that causes it.

Let's consider a special case. Just as in Maxwell's equations there are solutions that involve no sources, the same situation is true here. Consider the case either with no sources or in a region of space where they are too far away to play a role. Then, on the right-hand side of equation (36), we have zero.

When $T^{\mu\nu} = 0$, Einstein field equations simplify to

$$\mathcal{R}^{\mu\nu} = \frac{1}{2} \, g^{\mu\nu} \, \mathcal{R} \qquad (37)$$

Let's calculate $\mathcal{R}$ by contracting, on both sides, μ and ν. Since contraction is done on one upper and one lower index, we must first lower the ν index on both sides. This is done by multiplying with the appropriate version of the metric tensor. We get

$$\mathcal{R}^{\mu}_{\nu} = \frac{1}{2} \, g^{\mu}_{\nu} \, \mathcal{R} \qquad (38)$$

Then contraction of the left-hand side yields precisely $\mathcal{R}$. On the right-hand side, g^{μ}_{ν} is the Kronecker delta. Its contraction yields the number 4. We arrive at $\mathcal{R} = 2\mathcal{R}$. Therefore $\mathcal{R} = 0$.

We established that in the case of no sources, the curvature scalar is zero. Therefore in equation (36), where the right-hand side is already zero, we can also drop the $\mathcal{R}$ term on the left-hand side. We arrive at

$$\mathcal{R}^{\mu\nu} = 0 \qquad (39)$$

In the case when there are no sources, Einstein's equation has become simpler! It is called the *vacuum case*. In this case, the Ricci tensor is zero. As we have already pointed out, it doesn't imply that the Riemann curvature tensor be zero. There are nontrivial solutions to equation (39). Among them are gravitational waves with no sources, just like there can be electromagnetic waves even in space without electric charges.

There is also the Schwarzschild metric. It is roughly speaking analogous to a point mass. Outside the point mass there is no matter, no sources, nothing. Thus, just like Newton's equations, anywhere outside the mass, are the same as for empty space, the Schwarzschild metric, outside the singularity, is a solution to the vacuum Einstein field equation.[13]

---

[13]We use indifferently the singular or the plural when talking about Einstein field equation(s). It is one tensor equation, so it can be viewed as one or several equations.

It is conceptually a simple check, although the calculations are a real nuisance. You start from the Schwarzschild metric, sit down and spend the rest of the day calculating a ream of Christoffel symbols, get to the curvature tensor, contract it to arrive at the Ricci tensor, $\mathcal{R}^{\mu\nu}$, and observe that it is equal to zero. You can also do it by feeding the Schwarzschild metric into Mathematica on your computer.

Metrics for which the Ricci tensor is zero are called *Ricci flat*. Anywhere except at the singularity – where everything is crazy and meaningless anyway, just like for Newton's point masses – the Schwarzschild metric is Ricci flat. Let's repeat that Ricci flat is not the same as flat.

The Schwarzschild metric satisfies equation (36), except at the singularity. Gravitational waves satisfy it too. These are first basic facts about Einstein field equations. We will finish this volume of introduction to general relativity by studying gravitational waves in the next lecture. But first let's do a questions/answers session to answer specific questions from the audience.

Questions/Answers Session

Question: What happens if we do the contraction to calculate $\mathcal{R}$ not on the simplified equation (38), but directly on Einstein's equation where we leave $T^{\mu\nu}$?

Answer: Okay, let's do that. So we start from

$$\mathcal{R}_{\mu\nu} - \frac{1}{2} \, g_{\mu\nu} \, \mathcal{R} = 8\pi G \, T_{\mu\nu}$$

It doesn't matter if the indices are upstairs or downstairs. After multiplying by the appropriate version of g to move one of the indices upstairs, we contract the two indices. That gives us

$$\mathcal{R} - 2\mathcal{R} = 8\pi G \, T_\mu^\mu$$

T_μ^μ is the same thing as $T_{\mu\nu} \, g^{\mu\nu}$. Let's call it T. This scalar T is by definition what you get when you contract the two indices of

T_ν^μ. So we have

$$\mathcal{R} = -8\pi G \ T$$

Then we put this back into Einstein's equation. After minor re-organizing we get

$$\mathcal{R}_{\mu\nu} = 8\pi G \ \left(T_{\mu\nu} - \frac{1}{2} \ g_{\mu\nu}T \right) \qquad (40)$$

In other words, Einstein's equation can be written like equation (40) equating the Ricci tensor to something. But on the right-hand side you have to compensate by subtracting $\frac{1}{2} \ g_{\mu\nu}T$.

Q: Does T have an interpretation?

A: Yes. It is called the *trace* of the energy-momentum tensor. It is 0 for electromagnetic radiations, that is, for massless particles like photons or gravitons. For particles with mass, the trace of the energy-momentum tensor is not zero.

Q: You explained the interpretation of the Riemann tensor. Can you explain in the same way how to interpret the Ricci tensor and the curvature scalar?

A: We saw indeed that the Riemann tensor has to do with going around a little bump and seeing how much a vector, which locally doesn't change in each local flat ("best") coordinates, has rotated when we come back. I don't know any particular physical significance or geometric significance to the Ricci tensor or the curvature scalar. Whatever it is, it is not very transparent. They are much simpler objects than the full Riemann tensor. They average over directions. We lose information when we go from Riemann to Ricci. We can have $\mathcal{R}_{\mu\nu} = 0$ while the Riemann curvature tensor is not zero. An example that we will explore in the next and final lecture is gravitational waves (aka gravity waves).

Gravitational waves are comparable to electromagnetic waves: they don't require any sources to exist. Of course, in the real world you expect electromagnetic waves to be produced by an antenna or something. But as solutions of Maxwell's equations, you

can have electromagnetic waves that just propagate from infinity
to infinity like the Flying Dutchman.

In the same way, you can have gravitational waves that also have
no sources. Those waves satisfy $\mathcal{R}_{\mu\nu} = 0$. But they are certainly
not flat space. It will be somewhat satisfying to meet a geometry
that is Ricci flat but whose curvature tensor itself is not equal to 0.

This is of course also the case of the Schwarzschild metric as long
as you stay away from the singularity.

There is something in such space-time. There is real curvature,
tidal forces, all sorts of stuff. But $\mathcal{R}_{\mu\nu} = 0$. So $\mathcal{R}_{\mu\nu}$ has less
information in it than the curvature tensor.

Notice that it actually depends on the dimensionality. In four
dimensions there is less information in the Ricci tensor than in
the Riemann tensor. In three dimensions, it turns out that the
amount of information is the same. You can write one in terms of
the other. And in two dimensions, all of the information is in the
scalar. That is all there is. There is the scalar, and from it you
can make the other things.

Q: The information that gets lost when we go from the Riemann
tensor to the Ricci tensor does not affect the energy-momentum
tensor nor Einstein's equations. What is the meaning of this lost
information then?

A: It means that for a given source configuration, there can be
many solutions to Einstein's equations. They all have the same
right-hand side, namely $T^{\mu\nu}$. But they simply have different physical properties. For example the simplest case is to ask: what if
this energy-momentum stuff is zero?

If it is zero, does it mean that there is no gravitation, no interesting geometry at all? No. It allows gravitational waves.

Furthermore, select an energy momentum tensor and construct a
solution. Then you can add gravitational waves on top of it. It is
not exactly true, but it is roughly true that to any solution you

can always add gravitational waves. So the gravitational waves must be something that contains more information than just the Ricci tensor. And they do.

Q: Shouldn't Einstein's equations contain a cosmological constant?

A: We haven't said anything about the cosmological constant – whether it exists or not – because it can be thought of as part of $T^{\mu\nu}$. From this point of view, the "cosmological constant" is an extra tensor term on the right-hand side of equation (36). We could denote it

$$T^{\mu\nu}_{\text{cosmological}}$$

And that would not change the look of Einstein's equation. If it was indeed a scalar, we could write it

$$T^{\mu\nu}_{\text{cosmological}} = \Lambda \, g^{\mu\nu}$$

Then it would make sense to shift it on the left-hand side, because it would only be a part of the geometry. This yields the following equation frequently met in the literature:

$$\mathcal{R}^{\mu\nu} - \frac{1}{2} \, g^{\mu\nu} \, \mathcal{R} + \Lambda \, g^{\mu\nu} = 8\pi G \, T^{\mu\nu}$$

Here is a brief history of this added term. When Einstein wrote his equation, the common view among cosmologists was that the universe was the Milky Way, and that it was stable. But Einstein soon realized that his equation implied that the universe could not be stable. So he added an extra term to counter the instability.

Then, in 1929, Hubble[14] observed with instruments that the universe was expanding, as Friedmann[15] and Lemaître[16] had predicted theoretically some years before (see volume 5 of TTM on

---

[14] Edwin Hubble (1889–1953), American astronomer.

[15] Alexander Friedmann (1888–1925), Russian physicist and mathematician.

[16] Georges Lemaître (1894–1966), Belgian astronomer, physicist, and mathematician.

cosmology). Then the cosmological plug-in became irrelevant. Later Einstein called it "the biggest blunder of my career."

In fact, to this day, the cosmological constant is the subject of debate because it can frequently be called to the rescue as an "explanatory factor" in various riddles of the cosmos. However most theoretical physicists dislike ad hoc factors or procedures to solve (or fudge) a problem.[17]

Yet an ad hoc adaptation is sometimes *definitely* fruitful if one thinks of the introduction of quantas by Max Planck as an ad hoc sleight of hand to overcome a problem with black-body radiations.

Q: Did Einstein use the equation you derived in this lecture to explain the precession of the perihelion of the orbit of Mercury? And do you know of a document where we can see how this calculation was done?

A: It is certainly in one of Einstein's papers.

Let's talk about the two great successes that general relativity quickly achieved: the prediction of the bending of light rays from distant stars by the Sun (see the note on Arthur Eddington in lecture 7), and the explanation of the oddities in the orbit of Mercury.

Concerning the bending of light rays, Einstein did not have the Schwarzschild solution. But he did have the approximation to the solution at fairly large distances. Of course, the Sun is not a black hole. But outside the solar radius the geometry is exactly the same as that of Schwarzschild.

Far away from the radius, Einstein knew how to make a good approximation. The fact that the Sun is so big means that the corrections from Newton are small. They can be done using the

---

[17] Remember Ockham's principle, also called Ockham's razor, named after the Franciscan friar and scholastic philosopher William of Ockham (1285–1347): "*Pluralitas non est ponenda sine necessitate*." It can be translated as "One should not hypothesize complexity without necessity," that is, it is ill-advised to introduce additional parameters in a model just to fit the observational facts more easily.

technique of perturbation theory: you add some small corrections to something you already know, and fit them so that the equations still hold. For a light ray, you start from a nice straight line, do a little bit of perturbation theory, and work out the modification of the trajectory.

For the orbit of Mercury, most likely what he did was to start from the Keplerian orbits. That is, he started from Newton's solution. Then he fitted the small correction on the left-hand side to the small correction on the right-hand side. And he found exactly the right discrepancy in the perihelion of Mercury.

But then, within a few weeks after Einstein's first publication, Schwarzschild calculated his exact solution of the equation. And from them you could do much better than perturbation theory.

Let me mention one last point concerning the orbits of planets. The ordinary gravitational potential in $1/r$, leading to a centripetal force law in $1/r^2$, is pretty much the only law that leads, in the simplest case, to elliptical closed orbits without precession. You can see it as a curious accidental fact.[18] Or you can try to explain it with a theory.

In the next lecture, which ends the book, as an example of the exploitation of Einstein field equations, we will study gravitational waves, which are a particularly simple solution to those equations when the energy-momentum tensor is null, that is, when spacetime contains no mass, or equivalently no energy.

---

[18]"Accidental facts" are precisely the kind of facts that Einstein did not take for granted. It is the rejection of the accidental character of the equality between inertial mass and gravitational mass that ultimately led him to the theory of general relativity.

Lecture 10: Gravitational Waves

Lenny: *After the strenuous efforts of the last lessons, we will finish our conversation with easier stuff: the derivation of very simple solutions to Einstein field equations, which predict gravitational waves.*

Andy: *Those we begin to detect?*

Lenny: *Exactly.*

Andy: *Perhaps another chance to hear from aliens in the universe?*

Lenny: *Who knows? Enrico Fermi used to ask: But if they exist, they should already have visited us. Where are they?*

Andy: *Actually life can take so many forms that it is a narrow view to expect other brains similar to ours communicating with us.*

Lenny: *Yes, the best is to keep developing our own, while not disturbing too much the universe or even our planet.*

Introduction

The questions we want to address in this lecture are weak gravitational fields, linearity versus nonlinearity, and gravitational waves. We will finish with an overview of how Einstein field equations can be derived from the action principle.

Again working out the equations of general relativity is always unpleasant. We are not going to do it here. The calculations would fill pages even for simple things. And they probably would not be terribly illuminating. To learn the subject, you really have to sit down and compute yourself and solve the equations. On the other hand, the principles are straightforward. It is easy enough to explain what we get when we do solve equations. That is the way we will talk about gravitational waves: by writing down the equations and then writing their solutions.

To start with, we are interested in what could be called weak gravitational waves or fields.

Weak Gravitational Fields

We talk of a weak gravitational field when the gravitational variations over space-time are small enough that we can make the following approximation: the variations *squared* of the gravitational field can be taken to be zero.

When a quantity is small and we expand an equation involving it, the usual rule is to ignore terms that are of higher order than power 1 in that small quantity. The typical example is reducing the binomial theorem expansion to its first two terms:

$$(1 + \epsilon)^n \approx 1 + n\epsilon$$

One general technique to obtain new interesting solutions of an equation is to start from an equilibrium, then to consider small fluctuations, or perturbations, about the equilibrium situation.

In this lecture, we shall look at the simplest equilibrium solution of Einstein's equations. We write Einstein field equations (equations (36) of the last lecture) in the case without any matter – no energy-momentum tensor on the right-hand side. The equations of motion are then simply

$$\mathcal{R}_{\mu\nu} - \frac{1}{2}\, g_{\mu\nu}\, \mathcal{R} = 0 \qquad (1)$$

This can be simplified further. To do that let's see how we can figure out the value of $\mathcal{R}$. If we take the trace of both sides, after having raised one index upstairs, since $\mathcal{R}$ is the trace of the Ricci tensor $\mathcal{R}^\mu_\nu$, and 4 is the trace of g^μ_ν (which is the Kronecker-delta), we end up with $\mathcal{R} = 0$. The curvature scalar is zero.

Notice that the Riemann tensor doesn't have to be zero. But we can ignore the second term in equation (1). The equation becomes simply

$$\mathcal{R}_{\mu\nu} = 0 \qquad (2)$$

In fact, this equation won't really matter to us because we are not going to write down the details anyway. But it is Einstein's equation in a context when there is no energy-momentum tensor on the right-hand side of the full field equation.

What is an equilibrium situation? Generally speaking, it is a solution that has no time dependence. In our case it also doesn't have any matter on the right-hand side – matter being another word for the energy-momentum tensor.

Then there is only one equilibrium situation. It is just empty space with no time dependence. By such empty space we mean empty flat space,[1] no curvature, no interesting gravitational field. In that case, the metric $g_{\mu\nu}$ can have the simple $\eta_{\mu\nu}$ form we are familiar with.

Remember that it is improper to say "the metric is equal to such and such matrix," because the metric is a rank-2 tensor. In each coordinate system, it has a different expression with different components. For instance, in Euclidean geometry, if I wrote that the metric of a flat plane was just the Kronecker-delta symbol, you would correct me: "No, the metric of a flat plane is not the Kronecker-delta. What is true is that in an appropriate coordinate system it is possible to express the flat metric with the Kronecker-delta." Indeed, if I used other coordinates, the expression of the metric would not be the Kronecker-delta. We could use curved

---

[1] We specify empty *flat* space, because in this lecture we shall indeed find a solution to Einstein field equations that corresponds to empty space-time but not flat. It will have, however, a time dependence.

coordinates, polar coordinates, or any other kind of coordinates. Go back to figure 1 of lecture 2, if necessary, to refresh your memory. In other words, what is special about flat space in Euclidean geometry, or in Riemannian geometry, is that you can find coordinates in which the metric has a nice simple form.

The same is true in general relativity – in which, we have seen, the geometry is Minkowski–Einstein geometry. At each point X in space-time, the local metric has three positive and one negative eigenvalues. For flat space-time with no gravitational field, there is a choice of coordinates in which the metric $g_{\mu\nu}$ doesn't depend on X and has the simple form

$$\eta_{\mu\nu} = \begin{pmatrix} -1 & 0 & 0 & 0 \\ 0 & 1 & 0 & 0 \\ 0 & 0 & 1 & 0 \\ 0 & 0 & 0 & 1 \end{pmatrix} \tag{3}$$

The first row and the first column correspond to time. The four columns correspond to (t, x, y, z). Same for the four rows. This is the metric of flat space-time in the most appropriate coordinate system. It is an equilibrium solution, that is, it is a solution of the Einstein field equations, and this solution doesn't depend on time.

Remember that the complete description of motion is "the field tells the sources how to move" (they move along geodesics in space-time) and "the sources tell the field what to be" (it is a solution to Einstein field equations, which themselves generalize Poisson's equation of the Newtonian case; see equation (4) of lecture 9). In general the solution is time-dependent, of course. But here we are interested in an equilibrium solution, that is, independent of time. Secondly, we are interested in the equilibrium solution for the simplest type of source configuration, namely no sources.

We are looking at the case where Einstein field equations reduce to $\mathcal{R}_{\mu\nu} = 0$. They say that certain components of curvature are equal to zero. It is a purely geometric condition, and we have a solution that is particularly simple: no curvature at all anywhere. It is easy to check that it satisfies $\mathcal{R}_{\mu\nu} = 0$. This concludes the first step toward weak gravitational waves.

Let's turn now to a space-time that is close to empty space-time with no time dependency.

Suppose that, very far away from where we stand, some big phenomenon is happening. For example, a binary pulsar – which are two stars going around each other – is rotating rapidly and emits some complicated gravitational waves. Near the pulsar the gravitational field may be very strong. Even the gravitational waves might be rather strong. But if you go far enough away, the gravitational radiation, that is, the gravitational waves that are produced by this thing, are going to be very weak.

What does "weak" mean? It means that the metric can be chosen – again, we emphasize, in the appropriate coordinate system – to be equal to $\eta_{\mu\nu}$ plus something small. Small means that the components of the added term are much smaller than those of η. We write this

$$g_{\mu\nu} = \eta_{\mu\nu} + h_{\mu\nu} \qquad (4)$$

As far as I know, the added term is usually called h simply because it is the next letter after g.

Unlike $\eta_{\mu\nu}$, the perturbation tensor $h_{\mu\nu}$ is in general a function of position. When we say "position" we mean the position or event X with four coordinates in space-time, of course. The added small tensor $h_{\mu\nu}$ varies from place to place and also from time to time, and it might describe a wave.

We will come back to waves in a moment. But let's first ask ourselves: what are we going to do with this $h_{\mu\nu}$? Answer: we will take the metric of equation (4), calculate its Ricci tensor $\mathcal{R}_{\mu\nu}$, and set it equal to zero. That will give us an equation for the tensor h.

The equation that we actually get for h is not big enough to fill a whole page, but it is sufficiently big to be unpleasant. So I'm just going to be schematic. I'm going to show what goes into it.

First, let's look at the Ricci tensor $\mathcal{R}_{\mu\nu}$. It is really a combination of components of the Riemann curvature tensor $\mathcal{R}_{\mu\nu\tau}{}^{\sigma}$. I'm not going to write again the full curvature tensor (see equation (26)

of lecture 9), I'll just remind you what it contains, its structure.
The $\mathcal{R}$ shown below doesn't denote the curvature scalar, but the
full rank-4 curvature tensor. It has the following structure:

$$\mathcal{R} = \partial\Gamma + \Gamma\Gamma \qquad (5)$$

It contains first derivatives of the Christoffel symbols, and it contains the Christoffel symbols quadratically, that is, products of
two Christoffel symbols. I could put the indices everywhere, spell
out the various terms that look like $\partial\Gamma$ and the various terms that
look like $\Gamma\Gamma$. But expression (5) giving the structure is really all
we need.

What about the Christoffel symbol[2] Γ? Likewise its structure is

$$\Gamma = \frac{1}{2}\, g^{-1}\partial g \qquad (6)$$

The symbol g^{-1} means the inverse matrix of g, or equivalently
the metric tensor with two contravariant indices.

Just like g with covariant indices can be expanded as in equation (4), g^{-1} can also be expanded in powers of h. The first contribution to it is simply the inverse of the η symbol. It happens to
be η itself, because η is its own inverse. Then we have the same h
term but with a minus sign, just like $1/(1+\epsilon) \approx (1-\epsilon)$. So the
expansion of g^{-1} up to the first perturbation term is pleasantly
enough simply

$$g^{-1} = \eta - h \qquad (7)$$

where actually the same matrix h appears with a minus sign.

Now, dropping the factor $1/2$ because we are only looking at struc-
tures, equation (6) becomes

$$\Gamma \sim (\eta - h)\partial h \qquad (8)$$

---

[2]When we talk of "the" Christoffel symbol, it is really a collection of
objects indexed by three indices each running from 0 to 3; see lecture 3.
Since they display some symmetry, there are not quite 64 of them, though.

The wiggly equal sign $\sim$ is used to equate things that are not equal at all. :-)

Equation (8) is the *structure* of the Christoffel symbol. It contains one term that has the power one of h. And this is multiplied by the first derivative of h.

We are also going to assume that ∂h is small. This would make a mathematician jump. But physically it means that, far away from the pulsar, not only the ripple h itself is small, in the sense that it has a small amplitude, but its variations are attenuated too. So the derivative of h is also considered to be a small thing.

The term $\eta \partial h$ is one order of magnitude small. The term $h \partial h$ is two orders of magnitude small. It is quadratic in the fluctuation, or in the small gravitational field, so we ignore it. For instance, if $h = 0.01$ and $\partial h = 0.01$, then their product is 0.0001.

In this approximation procedure $\eta \partial h$ is just ∂h. From equation (8) we deduce that, in the approximation of weak gravitational radiation, the Christoffel symbol is simply proportional to some collection of derivatives of h.

What about the Riemann tensor then? It will contain derivatives of Γ and products of two Γ's. That will produce various kinds of $\partial^2 h$, and various products of the form $\partial h \partial h$, that is, second derivatives of the gravitational field and products of first derivatives of the gravitational field. Incidentally, h is called the *gravitational field*, or the *field of gravitational waves*. We have

$$\mathcal{R} = \partial^2 h + \partial h \, \partial h \qquad (9)$$

Immediately, we say that $\partial h \partial h$ is much smaller than $\partial^2 h$ and again we ignore it.

Finally the Ricci tensor is built by contracting the Riemann tensor. So whatever the devil the Ricci tensor $\mathcal{R}_{\mu\nu}$ is, it is composed of simple second derivatives of the metric tensor.

From that analysis we can conclude that Einstein's equations have a relatively simple form. There are still swarms of indices around;

the number of different terms involving second derivatives of $h_{\mu\nu}$ is significant. It is complicated enough, and we don't want to write it out here, but it is basically built out of second derivatives – second derivatives with respect to position, second derivatives with respect to time, and maybe even some terms combining a derivative with respect to position and a derivative with respect to time.

Furthermore there are several components of it. How many? Well, the Ricci tensor has a μ and a ν, each running from 0 to 3. That makes 16 components. But the whole works, whatever it is, is composed of second derivatives of h.

In summary, our equation of motion is some kind of equation that looks like this:

$$\partial^2 h = 0 \qquad (10)$$

This leads naturally to the next topic, namely waves.

Gravitational Waves

Equations like equation (10), especially in relativity, are usually wave equations. Let's recall what the wave equation for an ordinary wave looks like.[3] Let's say we look at a wave that is moving along the z-axis as time flows. So consider a wave field that we can call $\phi(t, z)$. It will satisfy

$$\frac{\partial^2 \phi}{\partial t^2} = \frac{\partial^2 \phi}{\partial z^2} \qquad (11)$$

This is the simplest wave equation that you can imagine. Simple solutions to it are sinusoidal waves that move either to the right or to the left, as in figure 1.

You can add different solutions, for instance a wave going to the right and a wave going to the left. It is still a solution of equation (11) because the equation is linear.

---

[3] The wave equation belongs to the great general equations of mathematical physics, along with Poisson's equation that we met in lecture 9 and the heat equation.

If we had more directions of space, the structure of the equation would be a little more complicated. Instead of just the derivative of ϕ with respect to z, we would also have derivatives with respect to x and to y. The equation could be rewritten

$$\frac{\partial^2 \phi}{\partial t^2} - \frac{\partial^2 \phi}{\partial x^2} - \frac{\partial^2 \phi}{\partial y^2} - \frac{\partial^2 \phi}{\partial z^2} = 0 \qquad (12)$$

There is an evident family similarity between the kind of equation like (10), and the kind of equation like (12). In fact by clever manipulation you can make these equations look exactly alike.

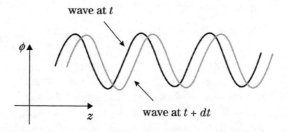

Figure 1: Simple wave $\phi(t,\ z)$ moving to the right.

Pay attention to the fact that equation (10) is not just one equation. There is one for each pair $\mu,\ \nu$. Let's put brackets around it to remind ourselves that $\mathcal{R}_{\mu\nu}$ has sixteen components. Equation (10) is rewritten

$$\left[\partial^2 h\right]_{\mu\nu} = 0 \qquad (13)$$

The components are not all independent. It is really quite a bit simpler than that. But still in principle there are 16 components, therefore 16 equations.

The collection of equations (13) is somewhat similar to Maxwell's equations in that respect. Maxwell's equations have the form of wave equations, for example for the electric and magnetic field. But there are several components. There are three components of the electric field and three components of the magnetic field. It sounds like there are only six equations, but in fact there are eight equations in the complete set of Maxwell's equations.

It is the same sort of pattern with equations (13): several equations, but all of similar form, and not all independent.

We wrote equations (13) only in a suggestive form. But we proved that they involve only second derivatives. In the process of producing equations (13), anything that involves h, or a derivative of h or a second derivative of h, is small, so *when it is multiplied by another small quantity* (as is the case when we do contraction, which introduces multiplicative terms, each of the form $\eta + h$), we omit the products of small terms because they become of a higher order of magnitude in smallness.

Remember that what we do is an approximation. But it is a well-defined approximation. Technically, it is the *linearization of Einstein's equations*. It simply means that in Einstein field equations we throw away everything of higher power than h. It is a good legitimate approximation when the gravitational radiation is weak. We reached equations of the form $\left[\partial^2 h\right]_{\mu\nu} = 0$, which we are not going to write down full fledged because they are messy.

Before we discuss their solutions, let's go back to an observation we made earlier – and in fact we repeated many times since lecture 2 on Riemannian geometry – that for the space-time to be flat, the metric doesn't have to be necessarily η. But with a proper choice of coordinates it *can be expressed using* η.

There are other ways to represent flat space-time. We can make a coordinate transformation on $\eta_{\mu\nu}$. The expression of the metric will change, as well as the form of the solution to Einstein field equations, but it will still be exactly the same physical solution.

Therefore there must be solutions of equations (13) that look like they are nontrivial. They will represent almost flat space-time: flat space-time with little ripples in it. Let's do that first in Euclidean geometry to clearly see this point.

Let's consider the flat page in Euclidean geometry. It is a flat space. So one of its metric's expressions is simply δ_{mn}, where

m stands for the X-axis and n for the Y-axis. Let's start with "good" Cartesian coordinates X and Y. Then we introduce coordinates X' and Y', which incorporate little disturbances. And let's express the coordinates (X, Y) in the coordinates (X', Y'):

$$X = X' + f(X', Y')$$
$$Y = Y' + g(X', Y') \tag{14}$$

This is a coordinate change. We haven't changed the space in any way. All we have done is change coordinates. The new slightly wriggly coordinates represented in the old "good" ones are shown in figure 2.

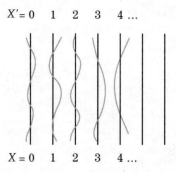

$X' = 0 \quad 1 \quad 2 \quad 3 \quad 4 \dots$

$X = 0 \quad 1 \quad 2 \quad 3 \quad 4 \dots$

Figure 2: Flat space with "good" coordinates (dark lines) and slightly wriggly coordinates (gray lines). We represented the X-coordinate; think also of the Y-coordinate, which is perpendicular, and its wriggly kin.

The metric in the "good" coordinates is

$$dS^2 = dX^2 + dY^2 \tag{15}$$

What is the metric in the (X', Y') coordinates? Remember: we are just in a Euclidean space, we are not doing space-time. So let's work out, just for fun, the metric in (X', Y'). We can write

$$dX = dX' + \frac{\partial f}{\partial X'^{\,m}} dX'^{\,m} \tag{16a}$$

and likewise

$$dY = dY' + \frac{\partial g}{\partial X'^{\,m}} dX'^{\,m} \tag{16b}$$

Our notations are a little awkward, because $X'^{\,m}$ represents the collection $X'^{\,1}$, $X'^{\,2}$, that is, simply X' and Y'. And equations (16a) and (16b) incorporate a sum with the summation convention.

Then we plug equations (16) into equation (15). We find out that dS^2 is not just $(dX')^2 + (dY')^2$. It contains $(dX')^2 + (dY')^2$ plus some cross-terms. If you work it out – assuming that f and g, and their derivatives, are small so that you can make the appropriate approximations – you will find that

$$dX^2 + dY^2 = (dX')^2 + (dY')^2 + h_{mn}\, dX'^{\,m}\, dX'^{\,n} \qquad (17)$$

The right-hand side is the square of the infinitesimal distance between two points calculated with the metric tensor expressed in the new wriggly coordinates.

Does this small correction to the metric tensor mean that the page is not flat anymore? Of course not. It just means that the new coordinates we use have wriggles in them (when we represent them in "good" coordinates; see figure 2).

It is a nice little exercise to compute, in equation (17), what the correction h is. You will get

$$h_{mn} = \frac{\partial f}{\partial Y'} + \frac{\partial g}{\partial X'} \qquad (18)$$

That will mean nothing to you until you try to work out an example, or until you try to prove it.

It is a small perturbation on the metric. It has the form that we already met in equation (4). But it doesn't represent anything physical. It is just a somewhat trivial change of coordinates.

In other words, there exist perturbations on the metric, which don't correspond to any physical effect on the space-time – it can remain flat even though the metric has the form of equation (4). Once again a flat space-time is one not where the metric is necessarily η but where *we can find* coordinates such that the metric in those coordinates becomes η.

Likewise, in equations (13), there are small perturbations we can write down that just represent curvilinear coordinates of space-time but don't change its geometry or its physics.

How can we eliminate those phony solutions that automatically solve the equation because they are just flat space-time, in curvilinear coordinates, and don't represent any real physics?

We do it by imposing more equations on the metric. I could write down the equations but it is not important. What is important is that they will wipe out the unwanted spurious solutions, the ambiguities on the coordinates, in other words the unphysical meaningless solutions of Einstein's equations. We can do that in a variety of ways.

Once that is done, the equations become pretty simple. They become wave equations. The components of h then satisfy perfectly ordinary wave equations, like equation (12). It becomes

$$\frac{\partial^2 h_{\mu\nu}}{\partial t^2} - \frac{\partial^2 h_{\mu\nu}}{\partial x^2} - \frac{\partial^2 h_{\mu\nu}}{\partial y^2} - \frac{\partial^2 h_{\mu\nu}}{\partial z^2} = 0 \qquad (19)$$

Each component of the fluctuation satisfies a wave equation. It means that all the components of the metric move down its axis like a wave, a linear wave.

On the other hand, there are also some constraints. There are more equations than one for each μ and ν. The reason is that we have these extra equations to eliminate spurious fake solutions, which we saw because of the coordinate ambiguity. Once we do that, we find out the physical solutions, the ones that really have meaning. We shall classify them.

Suppose we have a wave moving down the z-axis. Here is its equation:

$$\frac{\partial^2 h_{\mu\nu}}{\partial t^2} - \frac{\partial^2 h_{\mu\nu}}{\partial z^2} = 0 \qquad (20)$$

What does it look like? The simplest solution has the form

$$\phi = \sin k(z - t) \qquad (21)$$

It is a wave that at any fixed time t is a sinusoid; see figure 1. As time flows, it moves to the right down the z-axis with unit velocity. Unit velocity here means the speed of light. The parameter k is the frequency of the wave, also called the wave number. It is the number of complete oscillations per unit length. And of course, it can be any number. Short wave lengths have large k; long wave lengths have small k.

Each component of h will have a solution like that. It will be proportional to $\sin k(z - t)$. So let's write

$$h_{\mu\nu}(t,\ z) = h_{\mu\nu}^0 \sin k(z - t) \qquad (22)$$

The coefficient $h_{\mu\nu}^0$ is not a function of position. It is just a numerical coefficient that multiplies the sine of $k(t - z)$. That's about all that we can write down. Equation (22) is the nature of a gravitational wave, each component of the metric behaving like a wave moving down the axis.

However, as we said, there are more equations, to filter out spurious solutions. When we impose them all, we find out something interesting. To see it, let's first of all remember some facts about electromagnetism.

In electromagnetism we do the same thing: we solve Maxwell's field equations. We can solve them either for the vector potential or for the electric and magnetic fields, as in figure 3.

electric field in the horizontal $(x,\ z)$ plane

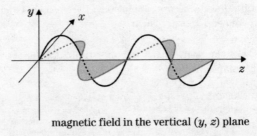

magnetic field in the vertical $(y,\ z)$ plane

Figure 3: Electric and magnetic fields photographed at an instant t.

When we write down all the equations, we find some constraints. The constraints are called the *transversality of the field*. It means that the electric and magnetic fields always point in directions perpendicular to the motion of the wave.

Not only the electric field and magnetic field solutions are waves, but they are transverse waves. That is, they are perpendicular to the axis of progression of the wave, which is the z-axis. And they are perpendicular to each other.

All the horizontal and vertical undulations shown in figure 3 vary over time in such a way that suggests a movement down the z-axis.[4]

Very similar things happen with gravitational waves. The consequence of the constraints, introduced to filter out spurious solutions, is that the waves have to be transverse.

To say that the waves are transverse means that the components $h_{\mu\nu}$ involving t or z are zero. The only components of h that are allowed to be nonzero are the components in the plane perpendicular to the direction of the wave.

If we incorporate this fact into the full set of equations – equations (13) plus the equations removing fake fluctuations – we find that a gravitational wave has a very simple form. The only nonzero components of the perturbation term added in the metric are

$$h_{ij}(t, \ z) = h_{ij}^0 \sin k(z - t) \tag{23}$$

where we use the dummy variables i and j, each of which taking its value in the set $\{x, \ y\}$.

Let's think of the spatial part of space-time at time t. And let's slice it into planes at different z values; see figure 4.

---

[4]When talking about transverse moving waves, usually nothing physically moves in the longitudinal direction. Think for instance of ripples moving on the surface of a pond. If you look at a cork floating somewhere, it will move up and down as the wave passes, but it won't move longitudinally.

The simplest example of a longitudinal wave, as opposed to transversal, is when we give a longitudinal bump to a long horizontal spring.

At a given location z and a given time t, the metric h_{ij} is the metric in one of the two-dimensional planes shown in figure 4. The components are simply numbers. In other words, at each z and t, the metric is just a set of numbers.

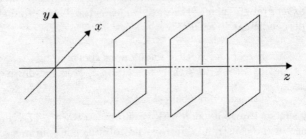

Figure 4: Spatial coordinates at a given time. The $x\,y$ planes at different z locations.

There is one more equation. It comes from Einstein's field equations. And it says that the trace of the metric h_{ij} is equal to zero. In equation form, this is

$$h_{xx} + h_{yy} = 0 \qquad (24)$$

Equations (23) and (24) form the whole set of equations. What it tells us is that the metric of a gravitational wave first of all satisfies

$$h_{00} = 0$$
$$h_{0x} = 0$$
$$h_{0y} = 0$$
$$h_{0z} = 0$$
$$h_{zx} = 0$$
$$h_{zy} = 0$$
$$h_{zz} = 0$$

That is to say, any component involving either t or z is equal to zero. The only components that are nonzero are

$$h_{xx}$$
$$h_{yy}$$
$$h_{xy}$$

They are given by equation (23). Furthermore, from equation (24), h_{xx} and h_{yy} are opposite to each other. The metric tensor is symmetric, so for any pair μ ν we have $h_{\mu\nu} = h_{\nu\mu}$.

That is pretty simple. But what does it mean?

Firstly, all of these components of the perturbation term vary with t and z, and only with t and z. If the wave is going down the z-axis, then by definition the variation is along the z-axis.

The quantity h_{xx} is equal to some coefficient h_{xx}^0 multiplied by the sine of $k(z - t)$. We can write the formulas for the three nonzero terms:

$$h_{xx} = h_{xx}^0 \sin k(z - t)$$
$$h_{yy} = h_{yy}^0 \sin k(z - t) = -h_{xx} \qquad (25)$$
$$h_{xy} = h_{xy}^0 \sin k(z - t) = h_{yx}$$

That $h_{yy} = -h_{xx}$ comes from equation (24).

Let's look at equations (25) at a fixed instant in time, for example time $t = 0$. As we move down the z-axis, there is a perturbation on the metric. The metric tensor is a little bit different from just a flat space metric – and it is a real curvature not just a coordinate effect. It oscillates as we go down the z-axis.

The components that oscillate are the components that have to do with the metric of the plane perpendicular to the wave.

What does it mean to have an h_{xx} added to η_{xx} – the latter being equal to 1 in the most natural coordinates? We write it

$$g_{xx} = 1 + h_{xx}^0 \sin k(z - t) \qquad (26)$$

It means that proper distances along the x-axis are a little bit different than what they would be if there was no perturbation term. Suppose we are talking about the distance between the point $x = -1/2$ and the point $x = +1/2$ (and $y = 0$). The proper distance is the *real distance* between the two points. Remember: the coordinates x's are just labels. The proper distance is given by the metric. And this proper distance may be either a little bit more than 1 meter or a little bit less than 1 meter (if we are working in meters). That depends on where we are on the z-axis.

For instance, at time $t = 0$, if we are at a point z where $h_{xx}^0 \sin kz$ is positive, that would mean that a stick of length 1 meter would be a little bit shorter than the distance between the two points. It would not be because the meter stick would be shorter – a meter stick is a meter stick – but because the plane would be stretched in the x direction.

Still at $t = 0$, what about the other spatial dimension? We have

$$g_{yy} = 1 + h_{yy}^0 \sin k(z - t) \qquad (27)$$

But h_{yy} is minus h_{xx}, from equation (24). We can write

$$g_{yy} = 1 - h_{xx}^0 \sin k(z - t) \qquad (28)$$

In other words, at a point z where the x direction is stretched, the y direction is compressed. A meter stick oriented along the y-axis would exceed a little bit the distance between the points with y-coordinates $1/2$ and $-1/2$. And conversely, when x is compressed, y is stretched. This is shown in figure 5.

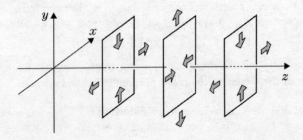

Figure 5: Alternative stretching and compression.

We can also reason at one fixed point z and see what happens over time. As time flows, the wave passes us. So alternatively our x-axis is stretched or compressed, and our y-axis is compressed or stretched.

This is a real physical effect. The gravitational wave creates curvature. The wave is, if you like, a ripple of curvature of space-time moving with the speed of light. At a given point on the z-axis, when the wave passes it, that is, when time flows, it creates a kind of tidal force there. The nature of the tidal force is to actually cause a meter stick in the plane at that position to be compressed and stretched. When it is compressed horizontally, it is stretched vertically, and when it is compressed vertically, it is stretched horizontally.

This is a measurable effect. Remember the 2000-mile man in figure 8 of lecture 1: he can feel the tidal forces, and with appropriate apparatus measure them. For instance, at a given point, if we took a piece of plywood, as in figure 6, we would see it alternatively being compressed horizontally and stretched vertically, and stretched horizontally and compressed vertically.

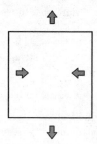

Figure 6: Tidal forces on a piece of plywood.

It is interesting to observe that there is also another solution. Then it is not h_{xx} and h_{yy} that are nonzero, but h_{xy}:

$$h_{xy} = h_{xy}^0 \sin k(z - t) \tag{29}$$

And h_{yx} is the opposite.

The metric would look like

$$\delta_{xy} + \begin{pmatrix} 0 & h_{xy} \\ -h_{xy} & 0 \end{pmatrix} \tag{30}$$

where δ_{xy} is the Kronecker-delta symbol, or equivalently the unit matrix. It still has $h_{xx} + h_{yy} = 0$, because both are zero.

It is the same phenomenon as in figure 6, except that the oscillating squeezing and stretching are along the 45° axes. It is not a new solution. It is just the initial solution rotated by 45°. Then any linear superposition of these two solutions is also a solution.

That is all the gravitational waves there are. You first pick the longitudinal direction in which the wave is moving, then you pick a set of perpendicular axes in the perpendicular plane, and you construct the wave alternatively stretching and compressing as described.

Questions/Answers Session

Question: How can we measure this squeezing and compressing?

Answer: We can use a strain gauge. The wave will create genuine stresses in the piece of plywood. A strain gauge will register them when the wave is going past.

If the wave were static instead of moving, then the piece of plywood would really be unaffected. Everything would be simultaneously squeezed, or stretched, the same way, the rulers that measure the plywood, the strain gauges that measure the stress, etc. But it is the oscillating character of the solution that really does create real honest stresses and strains in it. It does have real curvature oscillation.

Q: What triggers the gravitational wave?

A: In electromagnetism a moving charge, for example an electron rotating around a proton, or a current alternating in an antenna, creates electromagnetic waves. Similarly in gravitation, a moving mass will create a gravitational wave, for instance a star or a black hole or a pulsar or whatever, in orbit around another one.

The famous binary pulsar of Hulse[5] and Taylor[6] discovered in 1974, for instance, is made of two very concentrated masses, which rotate around each other at a rather small distance. Neither is a black hole. The whole thing is not a black hole. But it has a very strong varying gravitational field and it produces waves.

Near the binary pulsar, $g_{\mu\nu} = \eta_{\mu\nu} + h_{\mu\nu}$ is a bad approximation. The field is too strong to linearize the equation this way. But if you go some distance away, then the wave spreads out, dilutes itself, and gets weak enough that it becomes a good approximation.

If you are far away and take the z-axis as the radial line between you and the pulsar, you will feel a gravitational wave passing you, which has expression (26). There will be stresses and strains in the plane perpendicular to the line of sight to the pulsar.

Q: How can we detect deformation if the meter sticks are deformed too?

A: Meter sticks made of matter would indeed be affected. But a wooden meter stick and a steel meter stick would probably react differently. So in theory that is one way to detect the tidal forces created by the wave.

Another way to detect the gravitational wave is to create some system that has a resonance at the frequency of the gravitational wave. A system usually has its own natural frequency of oscillation. If it is reinforced by an oscillating force of the same frequency, bringing in some energy to compensate for that loss in the vibrations, it enters into a sustained or so-called driven resonance.[7] Under those circumstances the response will be particularly big.

---

[5]Russell A. Hulse (born 1950), American physicist.

[6]Joseph H. Taylor (born 1941), American astrophysicist.

[7]See for instance the famous example of the destruction of the Tacoma Narrows Bridge after it entered in resonance like a clarinet reed under the effect of wind: https://www.youtube.com/watch?v=3mclp9QmCGs

Notice that the gravitational waves we expect from various sources in astronomy are extremely weak. The additional terms $h_{\mu\nu}$ are extremely small, so small that the dimensionless effect on a rod would be something like 10^{-21}. The fact that people even contemplate measuring them is quite astonishing.

However they can be measured. That is the purpose of the Laser Interferometer Gravitational-Wave Observatory (LIGO) experiments conducted by Caltech and the Massachusetts Institute of Technology[8] in Livingston, Louisiana, and Hanford, Washington. LIGO is a gravitational wave detector. It is not a steel rod. It is a pair of mirrors, plus laser beams, like a modern version of the Fabry–Perot interferometer. An interference effect, due to the relative motion in the x and y directions, is produced and measured. But basically a gravitational detector is a system that is allowed to be set into oscillation.

The gravitational collapse of a black hole or the collision of two black holes happens once in a while in the universe. We can calculate how many such events, which are within range of the best imagined detectors, happen per unit of time. I think we could detect, in theory, one black holes collision per year. It is a lot.

The wave moves at the speed of light. Therefore, if we had at our disposal other means of detection of the black holes collision than the gravitational wave produced, we should see the signal at the same time as the wave.

Gravitational radiation is a pretty weak effect when you are far away. From the collision of two black holes, it is an enormous effect when you are close to it. It is then much bigger than any other kind of radiation that is emitted. But if the collision is taking place at large cosmological distances, it can be the case that the only way to detect it would be through gravitational radiation. That is indeed why LIGO is interesting. For astronomy it is a new instrument opening a new window on the universe:

---

[8]Waves were detected for the first time on September 14, 2015. See "Gravitational Waves Detected 100 Years After Einstein's Prediction," news release, LIGO, February 11, 2016, retrievable at https://www.ligo.caltech.edu/news/ligo20160211

Gravitational waves, which are vibrations of space-time, will enable us to detect and study events that took place before the time when the universe became transparent.

We can therefore look forward to making great progress in our knowledge of the beginnings of the universe (see volume 5 of TTM on cosmology).

Before the LIGO detection of 2015, gravitational waves had never been observed "directly."[9] But they had already been observed "indirectly" through other physical effects. Just like an orbiting system of electric charges emits electromagnetic radiations that carry off energy, masses that rotate about each other emit waves and lose energy. Their rotation will speed up a little bit due to this loss of energy.

The study of the binary pulsar of Hulse and Taylor showed a perfect match between the gravitational waves it should emit, the energy loss, and the rotation acceleration. This last effect can be measured by the change of frequency of the pulsating light received. Gravitational waves are the main phenomenon that causes the energy of the pulsar to decrease. The orbital period of the Hulse–Taylor binary pulsar is 7.75 hours. Since its discovery half a century ago, we have observed the decrease of its period exactly as predicted.

Q: The components h_{00}, h_{0x}, h_{0y}, h_{0z} are all zero. Does this mean there is no curvature along the time axis?

A: No. There is curvature in the time direction. The first derivatives $\partial h_{mn}/\partial t$ are not necessarily equal to zero. The curvature tensor has a whole bunch of products of such first derivatives and also has second derivatives. So it will have nonzero components involving time. But still it is a very special kind of curvature that is not generic.

---

[9] We put the word *directly* in quotes, because the concept of "direct" observation as opposed to "indirect" is in the view of some epistemologists, including Andy, a dubious distinction.

Q: Considering that Einstein's equations are nonlinear, how did we end up with solutions that are linear?

A: This is only because we assumed that the discrepancy between flat space-time and the actual (truly non-flat) solution to equation (1) was very small. The linearity is because the wave is weak. Generally speaking, small oscillations about an equilibrium, in a first approximation, can be taken to be linear. Remember the formula for the Taylor expansion:

$$f(x + h) = f(x) + hf'(x) + \frac{h^2}{2}f''(x) + \frac{h^3}{6}f'''(x) + \dots$$

If h is small we can take h^2 and the higher powers of h to be zero. Then we have

$$f(x + h) \approx f(x) + hf'(x)$$

That is the kind of approximation we make, for instance, when we assimilate a swinging pendulum (with oscillations of small amplitude) to a mass attached to a spring whose restoring force is an exact linear function of the displacement around the rest position. This is also the approximation we make when we linearize any type of vibration around a static equilibrium.

Q: How far off from an accurate solution are we? How much error do we make?

A: It depends how big is h. If h is $1/10$, and we want to add terms in h^2, the corrections would be 1%.

Q: If h is large enough that we have to add higher-order terms, will that affect the transversality of the waves?

A: No. The solutions will still be transverse, but won't simply add to give new solutions. For instance, two waves in opposite directions will no longer go through each other, they will scatter.

Q: If we compare these gravitational waves to mechanical waves, can we say that their speed depends on some stiffness coefficient?

A: The speed of gravitational waves is the speed of light. It is true that for mechanical waves propagating in some material, their speed depends upon the stiffness of the material, what is called its Young[10] modulus. So you can substitute stiffness of support for velocity of propagation. In that sense, that is, in that mechanistic way to understand physical phenomena, inherited from the sixteenth and seventeenth centuries, the medium in which light or gravitational waves propagate – what used to be called the *ether* – is the stiffest support possible.

In the old days, in the nineteenth century and before, people thought there was an ether. It was some sort of immaterial space, encompassing the whole universe, in which the notion of absolute location made sense. They thought that waves – for instance, light, after Young had shown it had wave-like properties – moved through the ether as mechanical waves move through a steel bar, or sound moves through the air. In other words, they thought that light was a vibration of the ether, which was propagating very fast. The very high speed of light was explained by the fact that the ether was very stiff. This mysterious substance was at the same time immaterial and the stiffest that existed. It had to have a very big Young coefficient because the speed of light is much higher than the speed of a wave propagating in a steel bar (which is itself about 20 times the speed of sound in air).

This way of thinking about the world, even though apparently sensible, led to all sorts of difficulties and subtle contradictions.

Anyway, it was definitively shattered by the Michelson–Morley experiments of the 1880s and by Einstein's explanation of 1905: the speed of light and all the laws of physics are the same in every inertial reference frame. Speeds don't exactly add up like they do in Newtonian physics. There is no ether. There is no such thing as an absolute motionless reference frame. We don't have an absolute position – let alone at the center of God's creation. Time is not universal, simultaneity is frame-dependent, etc. It took a while for the scientific community to accept it.

---

[10]Thomas Young (1773–1829), English polymath who made important contributions in optics (Young slits), solid mechanics, medicine, Egyptology, etc.

We all have in the mathematical toolbox we carry in our mind a 3D Euclidean space. But we cannot apply it, as would seem natural, to the entire universe. Most surprisingly, it is more like a ruler that we can use locally to measure things.

In recent history every century has brought new disturbing ways of thinking about the world. Let's mention the spherical Earth floating in space, the heliocentric theory, light travel that is not instantaneous, the relativity of inertial frames, etc. – to limit ourselves to a small sample and to stay in the realm of physics. One must be prepared for new scientific revolutions in every domain of knowledge, because scientific revolutions will not stop coming about. Curiously enough, another common sense idea – the idea that knowledge progresses toward a final universal truth – must also be relinquished.

Scientific revolutions are not new and will not stop, but – like in many other human affairs nowadays – one thing is new: their speed of renewal.

That is it for gravitational waves. You can learn more about them from any good book on general relativity. They will write down all the details of the solutions to the equations. These will fill a page or two, and won't be terribly illuminating. But we covered the basic principles.

Einstein–Hilbert Action for General Relativity

To finish this course, we want to present another way of thinking about the equations of general relativity. Again, solving the equations is too complicated, and we won't do it here.

Nevertheless, the basic idea is easy to grasp. It is to apply to general relativity the most central principle of physics that we have, namely the *principle of least action*. Indeed, all the systems that we know of, mechanical systems, electromagnetic radiation, quantum field theory, the standard model of particle physics, etc., are governed by an action principle.

In ordinary particle mechanics, the action attached to a trajectory
is an integral over time along the trajectory of the particle. We
met it in volume 1 of TTM. In field theory, the action attached
to a field is an integral over space and time. A field configuration
is a value of the field at every point of space and time. We met it
in volume 3.

We also saw that the first case is a special instance of the more
general second case. The trajectory of a particle can be seen as
a field whose substrate space-time is reduced to simply one time
axis, instead of a space-time with spatial and time coordinates;
see figure 7.

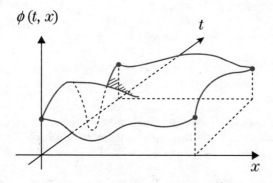

Figure 7: Field $\phi(t, x)$. If we omit the spatial dimension x, we have,
in the $\phi\, t$ plane, the trajectory of a particle over time.

Not every configuration of the field is a solution of the equations
of the theory, just like not every trajectory is a solution of the
equations of motion. Every trajectory is a thinkable trajectory,
but not every trajectory satisfies the equations of motion. In
the same way, every configuration or, better yet, every *history*,[11]
$\phi(t, x)$ is a thinkable possible value of the field in space and time.
But they are not all solutions.

---

[11]The variable t allows us to view $\phi(t, x)$ as the history of a spatial field
evolving over time, just like the trajectory of a particle can be viewed as the
history of its position.

In both cases, the behavior of the particle (its actual trajectory), or the behavior of the field (its actual configuration), is that which *minimizes the action*. Or more accurately, as we stressed many times, it is the configuration that makes the action stationary.

In the case of a particle, the quantity integrated over is called the *Lagrangian*. In the simplest situations, it is the kinetic energy minus the potential energy of the system. In the case of a field, the quantity integrated over is called the *Lagrange density* (because there is a spatial integration as well as an integration over time).

Let's focus on the case of a field with time and space in its substrate set, that is, $\phi(t, x)$. And x is a generic variable, which we could also denote X, standing for x, y, and z.

The action attached to this field has the form

$$A = \int dX \; dt \; \mathcal{L}(\phi, \; \phi_\mu) \tag{31}$$

where $\mathcal{L}$ is the Lagrange density. It depends on ϕ as well as on its derivatives with respect to the four coordinates X^μ (coordinate X^0 being then t). We denote these four derivatives ϕ_μ. Under the integral sign, dX stands for $dX^1 dX^2 dX^3$, that is, $dx\,dy\,dz$.

What do we do with this A? We want to minimize it, or more accurately to make it stationary. In other words, we want to find the field configuration that minimizes (we won't repeat each time "or make it stationary") the action subject to a constraint.

The constraint is that we are given the values of the field on the boundary of a region of space-time. Then the minimization process will yield the value of the field in the whole region. This region boundary plays the same role as the initial and final positions of a particle, at time t_1 and time t_2, in the case of particle motion. The solution of the minimization of the action yields, in that case, the whole particle trajectory between t_1 and t_2.

In the case of our field $\phi(t, x)$, or $\phi(t, X)$ if we have several spatial dimensions, the boundary can be the rectangular region in the $x\,t$ plane of figure 7, or it can be a more elaborate boundary.

The aim of this brief review of what we learned in volumes 1 and 3, about minimizing an action for a particle trajectory or for a field $\phi(t, X)$, is that the Einstein field equations can be derived from an action principle. It is a long computation but the action expressed by equation (31) is simple.

From the Lagrange density, we write the Euler–Lagrange partial differential equations in the multidimensional space-time case (see an example of such calculation in volume 3), which generalize

$$\frac{d}{dt}\frac{\partial \mathcal{L}}{\partial \dot{X}} = \frac{\partial \mathcal{L}}{\partial X} \tag{32}$$

And after a few hours of tedious but straightforward calculations, we land on the Einstein field equations.

Let's look at what the Lagrange density is. There are some interesting pieces in it. But first of all let's review what the concept of space-time volume is and how it is calculated.

Going one more step back, let's consider first an ordinary space, say a two-dimensional Riemannian manifold. The index m of the coordinates runs over 1 and 2, and so does n. When convenient, the first coordinate is simply denoted x and the second y. The manifold has a metric expressed by

$$dS^2 = g_{mn}\, dX^m\, dX^n \tag{33}$$

The square of the *distance* between two neighboring points is given by this formula (33).

Suppose now that we are interested in the *volume* of a small region of space, let's say a small rectangle with sides x to $x + dx$, y to $y + dy$; see figure 8. And suppose the metric has only two diagonal components g_{xx} and g_{yy}. We want to know the volume. Note on terminology: we use the generic term "volume," but of course in two dimensions, it is customarily called the area.

We would know it if we knew the actual proper distance associated with dx and the actual proper distance associated with dy.

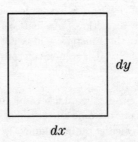

Figure 8: Small area, also called small "volume."

What is the proper distance associated with dx? It is given by equation (33): it is $\sqrt{g_{xx}}\; dx$.

So the area, or "volume," of the rectangle is

$$\text{volume} = \sqrt{g_{xx}\; g_{yy}}\; dx\; dy \tag{34}$$

The important point is that it is not just $dx\; dy$. No more so, mind you, than the proper distance between two neighboring points on the x-axis would be just dx. The coordinates are only labels. It is the metric that gives the distances – that is why it is called the metric. It warps everything if you will. Only in Euclidean geometry do the labels correspond directly to distances.[12]

We used the case where the metric only has diagonal terms, because it is easier to calculate the volume, and to illustrate the difference between $dx\; dy$ and the volume. In other words, equation (34) is just a special case when the metric has the form

$$g_{mn} = \begin{pmatrix} g_{xx} & 0 \\ 0 & g_{yy} \end{pmatrix} \tag{35}$$

---

[12]If you take a flat rectangular piece of rubber, print on it Euclidean coordinate axes, and then deform the piece of rubber with bumps, troughs, stretching, and compressing, the coordinate axes no longer give directly distances, but they remain useful labels to mark points on curvilinear axes. For the actual distances you will now need to know exactly the metric everywhere.

There could be, however, other off-diagonal components to the metric. Its general form is

$$g_{mn} = \begin{pmatrix} g_{xx} & g_{xy} \\ g_{yx} & g_{yy} \end{pmatrix} \tag{36}$$

In that case, the area, or "volume," is obtained from the *determinant* of the matrix expressed in (35). For a 2×2 matrix, the reader remembers that the determinant is $g_{xx}g_{yy} - g_{xy}g_{yx}$. For 3×3, it is a bit more complicated; it is given by Cramer's rule. For 4×4, it is obtained as a linear combination of 3×3, etc. But we don't need to know the formulas. The notation for the determinant is $|g|$. And the general formula for the volume in figure 8 is

$$\text{volume} = \sqrt{|g|}\, dx\, dy \tag{37}$$

This formula holds in any space, be it flat or curved.

If we consider a whole region on the 2D manifold, and we want to know its area, you guessed how we are going to calculate it: we integrate formula (37). It yields

$$\text{volume} = \int_{\text{2D region}} \sqrt{|g|}\, dx\, dy \tag{38}$$

If we are talking about the volume of a three-dimensional space, the formula is the straightforward generalization of equation (38), with a dz appearing as an extra differential element:

$$\text{volume} = \int_{\text{3D region}} \sqrt{|g|}\, dx\, dy\, dz \tag{39}$$

If we are in a four-dimensional space-time, it is useful to define a similar concept, which would now involve the fourth component dt as well. The rank-2 metric tensor is then expressed by a 4×4 matrix. In general relativity the metric is not Riemannian but Minkowskian, i.e., its signature is $- + ++$. Nevertheless at each point in space-time, it has a value and a determinant. And the volume can be defined generalizing again equation (39):

$$V = \int_{\text{4D region}} \sqrt{|g|}\ dt\ dx\ dy\ dz \qquad (40)$$

It is called the *space-time volume*, and it is an interesting quantity. *It is an invariant.* It is the same in every coordinate system. That is an important characteristic of the space-time volume. If you "re-coordinatize" the region, that is, if you change coordinates, in some sense the physical space-time volume of any region stays the same – which intuitively makes good sense.

Other quantities that will be invariant under change of coordinates are based on $|g|$. If we take, for instance, the volume element $\sqrt{|g|}\ dt\ dx\ dy\ dz$ and integrate it multiplied by any scalar $S(t, X)$, where X stands for $x\ y\ z$,

$$\int \sqrt{|g|}\ S(t,\ X)\ dt\ dX \qquad (41)$$

we also get an invariant quantity.

Indeed, it is just a sum of little volume elements, each multiplied by the value of a scalar field. Since the scalar is, by definition of a genuine physical scalar value, therefore an invariant too, the integrand in formula (40) is invariant. We deduce that expression (41) is invariant, even though writing it down explicitly necessitates the use of a set of coordinates.

In relativity theory – as in all theories – the action is always supposed to be the same in every coordinate system. Otherwise the laws of physics that came out of it would depend on the coordinate system.

If the laws of physics must be the same in every coordinate system, one way of ensuring that is to make the action an invariant. So it is natural to take for the action

$$\int dX^4 \sqrt{|g|}\ \times\ \text{some scalar depending on } X \qquad (42)$$

where dX^4 stands for the complete differential element.

Let's take Einstein's theory without any energy-momentum tensor. No sources, just a pure gravitational field, nothing else in space-time. What kind of scalar can we introduce in expression (42)? How many scalars are there that we could make up only out of the metric?

There is for instance the *curvature scalar*. Anything else? Answer: well, simply any number, 7, 4, or 16, or any number you like. So we can use for the scalar

$$S(t, \ X) = \mathcal{R} + \text{a number}$$

because numerical numbers are always regarded as invariant. If you see the number 7 at some point in space-time (not the component of something, but a pure number), everybody else sees the same number 7.

The number we add next to the curvature scalar depends on the laws of physics. It is itself a law of physics. It has a standard notation: it is denoted Λ (read lambda). And the formula for the action is

$$\int dX^4 \sqrt{|g|} \ (\mathcal{R} + \Lambda) \tag{43}$$

This additional number Λ is our controversial friend the *cosmological constant*. It creates a term in the Lagrange density that is proportional to the volume itself.

In the last lecture, we did not introduce the cosmological constant; we just discussed it in the questions/answers section that ended that lecture. If we don't introduce it, the action reduces to

$$\int dX^4 \sqrt{|g|} \ \mathcal{R} \tag{44}$$

Einstein field equations are what we obtain when we minimize expression (44), or make it stationary. Therefore at the solution we must have

$$\delta \int dX^4 \sqrt{|g|} \ \mathcal{R} = 0 \tag{45}$$

When we write δA, we mean the variation in A when we vary infinitesimally any variable upon which A depends.

Question: Where would the energy-momentum tensor appear in the Lagrange density $\sqrt{|g|}\ \mathcal{R}$?

Answer: The energy-momentum tensor would come in as additional terms in the Lagrange density that would depend on the material and other fields. For example there could be electromagnetic energy. There could be other kinds of energy. They would come in as additional terms in the integral in equation (45).

Equation (45) corresponds to the vacuum case. It is what governs gravitational waves.

Other things could indeed enter into the Lagrange density, to start with, as just said, sources of course. But they could also be made up from different fields. The Lagrange density, however, will always have the square root of determinant of g in factor. Its general form will be

$$\sqrt{|g|}\ (\mathcal{R} + \text{other terms}) \qquad (46)$$

For instance, recalling your course in electromagnetism, the electromagnetic field is described by an antisymmetric tensor $F_{\mu\nu}$. Expression (46) would then be

$$\sqrt{|g|}\ (\mathcal{R} + F_{\mu\nu}\ F^{\mu\nu}) \qquad (47)$$

That would govern the electromagnetic field.

But let's leave that out. And let's just finish with the Hilbert–Einstein action for the vacuum case given in equation (44), which we reproduced here:

$$\mathcal{A} = \int dX^4 \sqrt{|g|}\ \mathcal{R} \qquad (48)$$

It was discovered, I think, just about simultaneously by Einstein and Hilbert. But Einstein already had the field equations that he had derived following the method presented in lecture 9:

$$\mathcal{R}^{\mu\nu} - \frac{1}{2}\ g^{\mu\nu}\ \mathcal{R} = 8\pi G\ T^{\mu\nu} \qquad (49)$$

He already had the whole works and knew how it all fitted together.

Hilbert and Einstein independently sat down and asked themselves: is it possible to derive equations (49) from an action principle?

Both came up with the same answer that it is. They derived the field equations (49) from the action given by an extension of equation (48) when there are sources in space-time, that is, when elements from the energy-momentum tensor complete the Lagrange density.

In short, equation (45) is a physical principle that contains all the Einstein field equations.

To clarify further what we said in this section about fields and applying the principle of least action to them, let's ask ourselves the following question: what is the field $\phi(t,\ X)$ in general relativity, which appears in equation (31)? Answer: it is the metric

$$\phi(t,\ X) = g_{\mu\nu}(t,\ X) \tag{50}$$

Equation (48) describes an action that is made up out of a metric.

The *principle* to find the actual metric created by the physics is, as we have seen several times, to vary the metric a little bit in equation (48) and look for that which minimizes the action.

If we move away from the solution g, the action given by equation (48) should be somewhat bigger than at the solution.

The *mathematics* is to solve the Euler–Lagrange equation for a field

$$\sum_{\mu} \frac{\partial}{\partial X^{\mu}} \frac{\partial \mathcal{L}}{\partial \frac{\partial \phi}{\partial X^{\mu}}} = \frac{\partial \mathcal{L}}{\partial \phi} \tag{51}$$

where $\mathcal{L}(\phi,\ \phi_{\mu}) = \sqrt{|g|}\ \mathcal{R}$, and $\mathcal{R}$ is itself a function of g.

In conclusion, the Einstein–Hilbert form of general relativity is equation (45), reproduced here:

$$\delta \int dX^4 \sqrt{|g|}\, \mathcal{R} = 0 \qquad (52)$$

Steps to Derive the Field Equations from the Action

The following steps describe the complete procedure, starting from equation (52), to arrive at Einstein field equations:

1. Express the Christoffel symbols in terms of the metric. It involves first derivatives of the metric components.

2. Express the curvature tensor in terms of the Christoffel symbols. It involves first derivatives of the Christoffel symbols and products of Christoffel symbols.

3. Express the curvature scalar by contracting the curvature tensor.

4. Multiply that scalar by the square root of the determinant of the metric. The metric is a matrix of four rows and four columns. The determinant is a great big thing made up out of lots of components of the metric.

5. Apply the Euler–Lagrange variational equations (51) to the Lagrange density that you wrote.

After a few hours of calculations, you will end up with Einstein field equations.

This ends the book on general relativity. We hope that you enjoyed it. And we look forward to seeing you for the next books in The Theoretical Minimum series, which are *Cosmology* (volume 5) and *Statistical Mechanics* (volume 6).

Index

Here's what else you need to know to start doing physics!

The Theoretical Minimum: What You Need to Know to Start Doing Physics by Leonard Susskind and George Hrabovsky. A *New York Times* bestseller. World-famous physicist Leonard Susskind and hacker-scientist George Hrabovsky offer a first course in classical mechanics.

Quantum Mechanics: The Theoretical Minimum by Leonard Susskind and Art Friedman. Leonard Susskind and Art Friedman present the theory and mathematics of the strange world of quantum mechanics. The authors offer crystal-clear explanations of quantum states, entanglement, uncertainty, and more.

Special Relativity and Classical Field Theory: The Theoretical Minimum by Leonard Susskind and Art Friedman. Leonard Susskind and Art Friedman introduce readers to the physics and mathematics of light, with Einstein's special theory of relativity and Maxwell's classical field theory.